PUBLISHING AND THE ADVANCE
FROM SELFISH GENES TO GALI

For Sydney
With best wishes
from the author.

(It claims to be
an interesting
read...)

Michael
13.ii.14

PUBLISHING
AND THE
ADVANCEMENT OF
SCIENCE

*From Selfish Genes to
Galileo's Finger*

MICHAEL RODGERS

Imperial College Press

Published by

Imperial College Press
57 Shelton Street
Covent Garden
London WC2H 9HE

Distributed by

World Scientific Publishing Co. Pte. Ltd.
5 Toh Tuck Link, Singapore 596224
USA office: 27 Warren Street, Suite 401-402, Hackensack, NJ 07601
UK office: 57 Shelton Street, Covent Garden, London WC2H 9HE

British Library Cataloguing-in-Publication Data
A catalogue record for this book is available from the British Library.

PUBLISHING AND THE ADVANCEMENT OF SCIENCE
From Selfish Genes to Galileo's Finger

ISBN 978-1-78326-370-7
ISBN 978-1-78326-371-4 (pbk)

Designed and typeset by
Pete Russell, Faringdon, Oxon

Printed in Singapore by B & Jo Enterprise Pte Ltd

Dedicated to the Memory of
STEVEN PAULL RODGERS
1968–2010

FOREWORD

BY RICHARD DAWKINS

MICHAEL RODGERS has been an active and innovative player in the
world of science publishing for 35 years, working for five different
publishing houses. He was also a shrewd and far-sighted observer of
the whole publishing scene and of the public understanding of science
movement. His own successes as a commissioning editor are probably
unrivalled in the field, and his inside stories of these successes will be
of great interest to scientists and general readers alike. I don't doubt
that they will enjoy, as I did, the anecdotes and the personal impressions of the scientists, many of them very distinguished, whom he has
known and published.

Let me quote from my own memoir, 'Growing up in ethology',
my contribution in a volume of autobiographical chapters, *Leaders in
Animal Behavior: The Second Generation* (published in 2010). While I was
looking for a publisher for *The Selfish Gene* (in 1976), a colleague . . .

. . . introduced me to Michael Rodgers of Oxford University Press and
from then on there was never any question about who would publish it.
Michael simply had to have the book, and he would not rest until he got it:
'I haven't been able to sleep since I read it. *I must have that book.*'

There is something exhilarating in having your first book published,
especially when you have a publisher like Michael Rodgers. In joking
reference to ecological theory, he has described himself as a '*K*-selected
publisher' [see page 91], and that is exactly right, with the addition of the
adjective 'obstinate'. If your publisher is *K*-selected and obstinate, and if he
really likes your book, there are no lengths to which he will not go on its
behalf. One anecdote. The International Ethological Conference in 1977
was at Bielefeld in Germany. I was invited to give a plenary talk, and I used
it to introduce the idea of the extended phenotype as an outgrowth from
The Selfish Gene. The conference bookshop had a few copies of that book,

and they instantly sold out. The bookseller frantically telephoned Oxford University Press to try to get emergency reinforcements and she received a brush-off from one of the suits who infest such organizations: 'Whom do we have the honour of addressing? Well, you must understand, we have procedures to go through, you might get the books in three weeks if you are lucky.' Three weeks would have been much too late: the conference would be long over. The German bookseller appealed to me for help, and I telephoned Michael Rodgers. My memory still hears the slam of his fist on the desk: 'Good! You've come to the right man!' I don't know how he did it but, the very next day, a large box of books arrived in Bielefeld. I had indeed gone to the right man. Not just on that occasion, but in the first place. If you have a book to sell, go to a *K*-selected publisher. An obstinate one.

Michael Rodgers has spent a long career being a great editor and a staunch friend of science and scientific books. He has now turned author, and I think he deserves to be read.

CONTENTS

LIST OF ILLUSTRATIONS

ACKNOWLEDGEMENTS

Ivon Asquith, Susie Dent, and Lynette Owen read the opening chapters. They said they had enjoyed them, which boosted my confidence, but questioned whether enough readers outside the world of book publishing would find my account sufficiently interesting. In other words, could it really make commercial publishing sense? This usefully put a brake on my initial feelings of optimism, keeping me focused on what I very much wanted to achieve. I wanted to capture the interest of outsiders, those outside publishing, and even those outside science.

Sheila Lambie read each chapter as it was completed and provided useful comments. What I appreciated most were her regular helpings of encouragement. In 2010, Sheila asked me to give a talk to her postgraduate class of publishing students at Oxford Brookes University. Versions of my talk, 'Tales from Publishing: persuasion and cooperation', were subsequently given to her classes in 2011 and 2012.

Tim Rix read it and paid me a lovely compliment when he told me he had enjoyed it so much that as soon as he had finished, he went back to the start and read it all again. Tim's really important contribution began when he pointed out that I needed an additional chapter to open up my story for general readers, those outside publishing. The result was the present chapter 1. In content, it wasn't quite what Tim had had in mind when he made the suggestion but, he told me, he thought it nevertheless did exactly what was needed. Tim sadly died at the end of 2012.

I owe special debts of gratitude to Bernard Dixon and to Andrew MacLennan. Both read each chapter on completion, and both provided me with detailed lists of invaluable editorial comments and suggestions. Many passages were significantly improved as a result.

It is a pleasure to acknowledge the role of Alice Oven of Imperial College Press. For an author, it is wonderfully exhilarating to work with an editor who really believes in your book. Jacqueline Downs edited the text and her substantial number of requests for additional material resulted in a much improved final version.

PROLOGUE

I READ WITH HUGE ENJOYMENT Diana Athill's *Stet: An Editor's Life* (2000) and Jeremy Lewis's *Grub Street Irregular: Scenes from a Literary Life* (2008) shortly after they were published. Both are wonderfully written, both give outsiders an engrossing view of life in the world of book publishing. And in both cases the worlds described concern literature. As far as I know, no one has written about publishing from the point of view of science.

Might there be a very good reason for this? Might it be that, from the point of view of the general reader, science and scientists are not simply from another planet in terms of access, but also just plain dull? From my early days in publishing at the end of the 1960s I became convinced that science could be made just as absorbing to newcomers as subjects on the arts side. This wasn't simply a case of holding a view: it was something I found myself putting into practice as a central part of my first job in publishing, as science field editor working for the Oxford University Press. From the start, I knew that I had to make a special effort to make science not only accessible, the trivial bit, but also interesting, the interesting bit. In those days the accounts I wrote of my meetings with scientists were read by arts as well as by science colleagues and, later, when I presented books at sales conferences I knew that many of the sales reps I was trying to enthuse were wary of science. But I never found this a difficult task: science *is* fascinating and I enjoyed trying to communicate this. I discovered something else, too, that I enjoyed. Whenever I learned something new, a grand idea or a small but pleasing detail about science, or an amusing anecdote, my automatic reaction was to want to relate it to others. I stumbled into publishing and found out by chance that it enabled me to do the things I most enjoyed doing.

Working with Richard Dawkins on *The Selfish Gene* came early on in my career as a commissioning editor, and with Peter Atkins on *Galileo's Finger* towards its end. Their books struck me as being exactly right to bring together to use in this volume's title.

One delightful phrase stood out for me when I began thinking about a subtitle: 'the life it brings'. I first saw it when collaborating with Walter Gratzer. In the late-1980s I commissioned his anthology of writing about science, *The Longman Literary Companion to Science* (in America, Norton's *A Literary Companion to Science*). Reading his collection resulted in my searching for many a book, both fiction and non-fiction, I'd never previously come across: my appetite whetted by an extract, I absolutely had to read the whole thing. One of these was Jeremy Bernstein's *The Life It Brings: one physicist's beginnings*, a collection of beautifully written stories from physics. It was out of print by the time I wanted to read it but I spotted a copy (price, $6) in the amazing second-hand book emporium in New York, Strand Books, during a visit there in the mid-nineties. At the end of his Preface, Bernstein tells us about a letter written by the physicist Robert Oppenheimer to his younger brother, Frank, in the 1930s, quoting its conclusion: 'I take it that Cambridge has been right for you, and that physics has gotten now very much under your skin, physics and the obvious excellences of the life it brings.' 'What a wonderful phrase', writes Bernstein: 'the obvious excellences of the life it brings.' For me it perfectly encapsulates what science book publishing brought for me: meeting and working with extraordinary people, and discovering at first hand ideas that, when hearing about them for the first time, were simply breathtaking.

For that reason, 'Publishing and the life it brings' seemed to be the perfect subtitle. But Tim Rix thought it was perhaps too personal, suggesting that I should have one that was more explicitly descriptive, even if more prosaic. This prompted me to come up with 'Publishing and the advancement of science.' Later, Alice Oven told me that my proposed title and subtitle would work better if they changed places with one another. They were both right, of course . . .

CHAPTER ONE

HAWKING, EINSTEIN
AND
POPULAR SCIENCE

A Brief History of Time by Stephen Hawking was published in 1988 and it famously became a publishing phenomenon, going on to sell over ten million copies. It became a part of popular culture, with references to it in mainstream films, including *Legally Blonde, Donnie Darko* (both 2001) and *Harry Potter and the Prisoner of Azkaban* (2004), and in an episode of the television serial *Lost* (episode 7 of series 3, in 2007). Yet, also famously, the book has had many fewer readers than buyers. The outsider might be forgiven for finding all of this wholly bemusing. What was going on?

It was in the mid-1970s that I first read about Stephen Hawking, a brilliant physicist at Cambridge, confined to a wheelchair on account of motor neurone disease. At that time I was a commissioning editor at the Oxford University Press, and one of my special interests was popular science. A few years later, in the autumn of 1978, the publisher Robin Denniston joined the OUP to take charge of its academic publishing division. Robin's interests were wide and he was soon circulating notes to the division's editors, suggesting ideas for potential authors who might be approached about writing for the OUP. In one note sent to me Robin wondered if it might be worth contacting Stephen Hawking. With no hindsight to guide me, I decided not to follow up the suggestion, thinking it extremely unlikely that the high profile, busy Hawking would have any interest, even if he had the time, in writing a popular account of his subject.

Ten years later, at the beginning of June 1988, I happened to be visiting New York a few weeks after the publication there of *A Brief History of Time* by Bantam Books. The hotel where I was staying had placed in my room a copy of the latest issue of the *New Yorker* magazine and, with an hour or so to spare on my first afternoon, I casually leafed through its pages. I came to the book reviews section and suddenly became engrossed: by chance I had come upon a wonderful review of the book by the physicist and writer, Jeremy Bernstein. Intrigued, I bought a copy of the book that same day.

This timing proved to be a stroke of luck. The copy I bought was from an early printing and this meant that it included a short Introduction by Carl Sagan. Sagan held the copyright, permission had not been given to use the piece in all future printings, and eventually it was withdrawn. In his Introduction, Sagan relates that in the spring of 1974 he was at a meeting in England, sponsored by the Royal Society of London, to explore the question of how to search for extraterrestrial life. Sagan went on: 'During a coffee break I noticed that a much larger meeting was being held in an adjacent hall, which out of curiosity I entered. I soon realized that I was witnessing an ancient rite, the investiture of new fellows into the Royal Society, one of the most ancient scholarly organizations on the planet. In the front row a young man in a wheelchair was, very slowly, signing his name in a book that bore on its earliest pages the signature of Isaac Newton. When at last he finished, there was a stirring ovation. Stephen Hawking was a legend even then.'

It was another three years before I finally read my copy of the book. This followed a request from John Durant, the UK's first Professor of the Public Understanding of Science, based at the Science Museum Library and the adjoining Imperial College in London. John was editor of a new quarterly journal, *Public Understanding of Science*, about to be launched at the beginning of 1992, and he asked me to write an essay review on 'The Hawking phenomenon', to be ready for the journal's second issue (April 1992). My brief was to explain the astonishing sales of *A Brief History of Time*, and assess its impact on popular science publishing.

My starting point was the fact that lectures on fundamental physics and cosmology at annual meetings of the British Association for the Advancement of Science continued to attract packed audiences (I had attended many of them over the preceding couple of decades) and that books on serious popular physics continued to sell. Book publishers with popular science lists had long recognized particle physics and cosmology as good areas: they were exciting subjects with a powerful appeal to the imagination.

The best-known example of the genre to appear in the 1970s was Steven Weinberg's account of the early universe, *The First Three Minutes*, published in 1977. It sold well, and when *A Brief History of Time* was published, eleven years later, anyone familiar with Weinberg's book would probably have felt reasonably confident in predicting a respectable sales performance at least: like the earlier book it had a great title, a straightforward writing style, and an author who was a leading light in the field. But the sales of Hawking's book very quickly reached spectacular levels. Why?

At the outset, I argued, the book had three things going for it. First, it had a brilliant title. Distinctive titles really are important and I listed three especially memorable ones that readily sprang to mind: Weinberg's *The First Three Minutes*, Richard Dawkins's *The Blind Watchmaker*, and Roger Penrose's *The Emperor's New Mind*. Hawking had also added an enticing subtitle, *From the Big Bang to Black Holes*. Second, it promised a great deal: an account (according to the jacket blurb of the American edition) 'for those of us who prefer words to equations' of the origin, nature, and ultimate fate of the universe, all in fewer than 200 pages. Finally, it was written by an insider of considerable stature. The best popular science books are, in my view, written by participants, for only they can convey a genuine insight into the insider's world: what it feels like to be there.

Popularization is about more than simply making science understandable and, to illustrate this point, I quoted John Brockman, an American literary agent who has specialized in science (and specialized too in securing large royalty advances): 'In the old days, we had journalists writing up and scientists writing down. Now educated people want the real stuff from the horse's mouth.'

The three factors I had listed were doubtless sufficient to start the ball rolling, I concluded, but the 'Hawking effect', a public imagination captured by the thought of a brilliant mind imprisoned in a paralysed body, unravelling the secrets of the universe, must have come into play fairly swiftly, because the first printing of the American edition of 40,000 copies sold out so quickly that the publishers were taken by surprise.

I turned to the widely reported fact that the great majority of readers who had started to read the book had soon thrown in the towel. The frontiers of modern cosmology involve difficult ideas and I confessed that I had certainly not found *A Brief History of Time* an easy read. That point made, a good popular science book *should* be stretching. The trouble with this one was that a number of tough concepts, which were vital for following the argument, were explained at too fast a pace for general readers lacking a background in physics. After publication, Hawking himself identified two concepts which with hindsight he realized should have been explained with greater care: Feynman's 'sum over histories', and imaginary time. John Maddox, then editor of *Nature*, singled out particle spin as another. I am including these abstruse terms here to bring home the scale of the challenge facing Hawking in his desire to popularize this fascinating but difficult subject.

Identifying difficult passages in a draft typescript is easy enough for a publisher's editor, representing the lay reader, to do; what is not so easy is deciding on how far to go in expanding explanations and slowing the pace. Publishers of popular science talk about a general readership, but what does this mean? Scientists from other disciplines? Humanities graduates? Or, as Hawking himself was reported to have wanted to reach, plumbers and butchers, as well as doctors, lawyers, and science students? Only painstakingly developed course textbooks for science students are tuned to be read by precisely defined readerships, with their precisely defined prerequisite knowledge bases. With popular science, it is not so straightforward since a book written as a lay reader's textbook is unlikely to be very appealing. Many well-written popularizations of demanding subjects can be appreciated at different levels, with professionals in the subject drinking deeply, and outsiders experiencing something of the intellectual thrill of being there, without having to follow every detail. My own favourite example of a book where this was successfully achieved is Richard Dawkins's *The Extended Phenotype*.

In *A Brief History of Time*, Stephen Hawking may not have succeeded in making himself completely understood by every reader he wanted to reach, but if this was inevitable he does at least hold out hope for their children, or grandchildren. Towards the end of the book he writes:

> In Newton's time it was possible for an educated person to have a grasp of the whole of human knowledge, at least in outline. But since then, the pace of the development of science has made this impossible. Because theories are always being changed to account for new observations, they are never properly digested or simplified so that ordinary people can understand them. You have to be a specialist, and even then, you can only hope to have a proper grasp of a small proportion of the scientific theories. Further, the rate of progress is so rapid that what one learns at school or university is always a bit out of date. Only a few people can keep up with the rapidly advancing frontier of knowledge, and they have to devote their whole time to it and specialize in a small area. The rest of the population has little idea of the advances that are being made or the excitement they are generating. Seventy years ago, if Eddington is to be believed, only two people understood the general theory of relativity. Nowadays tens of thousands of university graduates do, and many millions of people are at least familiar with the idea. If a complete unified theory was discovered, it would only be a matter of time before it was digested and simplified in the same way and taught in schools, at least in outline.

Does it matter that most readers who at least opened *A Brief History of Time* gave up after twenty, thirty or forty pages? My view was, and still is, a positive one. Rather than forming a lasting impression of the impenetrability of modern science, vowing never again to pick up a book about science, some readers who gave up would surely have become aware that science is a human activity, that science is not dry and alien, progressing inexorably without touching ordinary people, and that scientists, even of Hawking's stature, make mistakes.

Finally, for my piece for *Public Understanding of Science*, I turned to the question John Durant had expressly asked me to address: how had the extraordinary impact that *A Brief History of Time* previously had affected the popular science book-publishing world? An immediate point to make was that no one should have been surprised to see a crop of titles written 'for people willing to admit they didn't quite understand *A Brief History of Time*'. The description was from a publicist at Harvard University Press, speaking about their *Ancient Light: Our Changing View of the Universe* by Alan Lightman (published in 1991).

Two other obvious consequences, given the enormous sales achieved by Hawking, 'are the ease with which publishers can now be persuaded to sign up proposals for books by the giants in science, and the size of the royalty advances which can be obtained for them'. This was certainly true at the time, the early 1990s, though the trend towards ever greater royalty advances inevitably came to an end, with a number of publishers discovering that the sales of some books expensively acquired were insufficient to earn back their lavish advances.

The examples I cited give a taste of the frenzied activity at the time. At the Frankfurt Book Fair in 1990, the American agent John Brockman sold a book to be written by the Nobel-laureate physicist Murray Gell-Mann for $550,000 (to Bantam, Hawking's publisher) on the basis of a 32-page synopsis; the advance topped one million dollars with the sale of foreign rights. In January 1991, Bantam snapped up *The Inflationary Universe* by Alan Guth on the basis of an eight-page synopsis sent by fax. I quoted Brockman's reported comment: 'Seven or eight years ago, you could take an eminent science book and sell it to an academic press for $5,000 and to Germany for $1,500, and you'd be thrilled. Today, you can take an eminent scientist . . . and within 12 hours have $225,000 and then auction the book to Germany for 80% of that. By the end of the week, with foreign sales, you will have three quarters of a million dollars.'

Note, as I did in my piece, that all of the action seemed to be in the United States. Brockman had mentioned Germany but not the UK.

I went on:

> When Hawking was initially considering writing a popular book he thought about doing it with Cambridge University Press but eventually decided on an American house because, he said, he wanted the book to sell in huge numbers. When his American agent offered the proposal to publishers in the US (in 1984) he had no problem in selling it: 'In New York practically every publisher in town was interested in the book. I had six-figure offers from about six . . . '. Trying to sell it in the UK was less easy: 'British publishers were the most sceptical I encountered. In Italy and Germany, for example, I sold the book blind for advances of about $25,000 to $30,000.' After receiving disappointing offers from some British publishers he withdrew, concluding that they were not serious enough. The UK rights were sold to Bantam in London some three years later.

I started the conclusion of my piece on something of a downbeat note, but ended on one of optimism: 'From my point of view, as a commissioning editor of popular science books in Britain, this is all very depressing. Certainly British publishing houses have successfully published numerous popular science books of quality, but there is nothing like the same excitement here, amongst both publishers and readers, as that to be found in America. It would be nice to think that the Hawking phenomenon might begin to breathe some life into British popular science publishing and book buying, even if only slowly.'

One factor contributing in no small way to the lower status enjoyed by popular science books in the UK at that time, especially compared to America, was the reluctance of literary editors of newspapers and periodicals in Britain to take science and its popularization seriously. How important was this in accounting for my complaint that, for popular science, there was nothing like the same excitement in the UK as there was in America, or indeed in Germany?

Let me start by returning to *A Brief History of Time* and what soon became a widely reported decision of Hawking's, this driven by his determination to reach as large a readership as possible. In his Acknowledgements section, Hawking writes: 'Someone told me that each equation I included in the book would halve the sales. I therefore resolved not to have any equations at all. In the end, however, I *did* put in one equation, Einstein's famous equation, $E = mc^2$. I hope that this will not scare off half of my potential readers.'

Hawking was on pretty safe ground. The equation is certainly famous: it is probably the only example of a scientific equation instantly recognizable by a sizeable proportion of the public at large. And an even greater proportion will surely have heard of Einstein himself. This widespread fame is centrally relevant to what follows.

In 1982, the Oxford University Press published *Subtle is the Lord: The Science and the Life of Albert Einstein*, by Abraham Pais. The Dutch-born American physicist and science historian had known Einstein personally—they had been colleagues at Princeton's Institute for Advanced Study—and his biography was praised as an important and superb piece of scholarship. The hardback edition sold over 30,000 copies, with a subsequent paperback selling a similar quantity. Of the hardbacks, more than 25,000 were sold in the United States, the remainder throughout the rest of the world, about 2,500 of them in Britain.

The story of this publishing success was told a few years later in an article by John Manger, published in the journal *Nature*. Manger, then editorial director for science and medicine at the OUP, was the commissioning editor originally responsible for signing up Pais's biography. The main purpose of his piece was to comment on how the book was treated by the respective review media in Britain and in the US.

On both sides of the Atlantic, science magazines and periodicals devoted largely or completely to reviews did well by the book. In the US it was reviewed by the *New York Review of Books*, *Science*, *Physics Today*, *Scientific American*, and numerous other weeklies and journals; in Britain by the *Times Literary Supplement*, *Times Higher Education Supplement* (now the *Times Higher Education*), *Nature*, and *New Scientist*.

But Manger's interest was in focusing on how the review coverage by the general media differed in America and Britain. In the US, the book attracted wide attention, from the *New York Times*, *Los Angeles Times*, and *Christian Science Monitor* to some thirty newspapers across the country (Manger's two examples, the Youngstown [Ohio] *Vindicator* and the Phoenix [Arizona] *Republic*, give an immediate feeling for just how local and specialized was the coverage achieved). The review in the *New York Times* was the lead, with an accompanying picture, in the Sunday books section, perhaps the prime review position in the world. Bookstores enthusiastically organized window displays of the book, and a science book club took several thousand copies and then came back for more.

None of this publicity was particularly exceptional for an American trade publishing house, Manger tells us. What *was* exceptional, he goes on, was the willingness of review editors in such a wide variety of newspapers and magazines to devote space to such a book. After all, writes Manger, it was 'not a simple layperson's guide to Einstein, but a scholarly, yet readable, account of the man and his not very simple science'.

The book's reception by the general media in Britain was in stark contrast. The approaches by the Oxford University Press to the national weeklies and dailies for review coverage were, in general, met with 'bewilderment':

> "Science? We don't have anyone who can review science books"; or "Our readers aren't interested in science"; or "We have 20,000 books a year submitted to us for review, we can't cover them all, you know." The only national newspaper review appeared in the *Observer*, seventeen column inches of it. Otherwise, except for BBC coverage in *Kaleidoscope* and on the World Service, the book was virtually ignored.

For Manger, the experience prompted two questions. Is the British public not interested in science books? Or is it being poorly served by the review media? A revealing—and amusing—story is provided to help convince those who might not yet have been convinced that the answer to the second question was a resounding 'yes'. One response to the OUP's attempts

> to attract attention to Pais's book was "Well you'd expect the Americans to be interested in Einstein, but he's not really very big over here." Who is big over here? In the week we would have expected reviews of *Subtle is the Lord* the dailies and weeklies were obsessed (in terms of review space) by a volume of the diaries of Beatrice Webb. Many column inches, plus suitable period photographs, were devoted to the earnest, but ultimately

trivial, meanderings of this fine Fabian. In the great scheme of things, it is difficult to believe that Einstein is not a more important figure to more people, even in Britain, than Beatrice Webb.

A literary editor would doubtless discount this as a somewhat tendentious conclusion, Manger continued, with the editor retorting that these kinds of judgements cannot and should not be made when it comes to deciding which books are reviewed in the national press. The literary editor must do what is best for circulation—quite right. 'But is it a valid assumption that 99 per cent of the readers of the daily and weekly papers have no interest in science for 99 per cent of the time? Because that is the legitimate conclusion to be drawn from the amount of review space given to science books.'

Per head of population, Pais's biography of Einstein sold twice as well in the United States as it did in Britain. But it is nonetheless almost certainly true that there is a greater interest in science in Britain than the media lead themselves to believe, was Manger's conclusion. 'The example of Pais's book demonstrates that science books do not have to be trivial to be saleable in global terms. The pity of it is that the review media in Britain remain stubbornly resistant to educating themselves or their readers.'

Manger's piece for *Nature* was spot on, though I doubt that it had much, if any, effect on the thinking of literary editors in the UK on how they perceived the significance of popular science, or the potential appetite for it. But my guess is that the publication of *A Brief History of Time*, with its huge sales, did more than anything else at last to begin to change attitudes. Something else too had changed by then: it was no longer felt to be acceptable to believe it smart, or amusing, to admit to having a complete ignorance of science.

Let me draw this chapter to a close with a delightful anecdote linking Einstein and popular science. J. G. Crowther, the world's first scientific journalist (appointed in 1928 to write for the *Manchester Guardian* by the paper's renowned editor, C. P. Scott: see Chapter 2), describes in his memoir, *Fifty Years with Science* (published in 1970), a visit to Berlin in 1930. There he met F. A. Voigt, an Englishman who was Berlin

Correspondent of the *Manchester Guardian*. Voigt told Crowther that he had once received instructions from Scott to see Einstein and commission him to write an article for the paper. Crowther goes on: 'He was to offer him a fee of, I think, five guineas. He called on Einstein, and had some difficulty in persuading him to write for the press, because he had not done so before. He made no comment on the fee, and was apparently not interested in it. After a week or two, the article was produced...'

The piece was on Newton, published on the 200th anniversary of his death, this explained in an editorial introduction: 'We are glad to be able to offer our readers an appreciation by the greatest physicist of the twentieth century of his great predecessor Newton, who died on March 20th, 1727. Dr Einstein had intended to write on Newton only in a German scientific journal, but he has kindly consented to write an article also for the *Manchester Guardian*.'

Einstein's commissioned piece, published by the paper on Saturday, 19 March 1927, was substantial: 3,000 words, occupying some sixty column inches.

Crowther's account continues: 'Scott was pleased with it, and sent a cheque of double the agreed amount to Einstein.' (He had been surprised, it seems, with the length of the submitted article, this being considerably longer than expected.) Crowther ended his story: 'Voigt heard afterwards that Einstein carried the cheque around with him for some time, showing it to his friends and saying that it was the first time in his life that he had ever been paid more than had been agreed.'

Back to Hawking, I had no involvement with *A Brief History of Time*, the most successful popular science book—in terms of sales—ever published. Nor did I have anything to do with Pais's biography of the most famous scientist in the world. My involvement with science publishing was somewhere in the middle of those two extremes. But what an extraordinary middle it proved to be: huge, and rich, and deep.

What follows is an account of one commissioning editor's experiences whilst working in that vast, fertile middle.

CHAPTER TWO

DISCOVERING
THE WORLD OF SCIENCE
AND
SCIENTISTS

My BACKGROUND before publishing was in science. At university I studied chemistry: research for a doctorate after my first degree, followed by a year in America on a postdoctoral fellowship. On return to the UK I tried two different sorts of career. The first was as a research chemist in industry, but this lasted less than a year when I finally realized that research wasn't really for me. And the other was as a chemistry teacher at a grammar school. This second attempt to find the right job initially worked well—I much enjoyed teaching—but the period when I felt I might have found the right occupation ended after a couple of years, and I later worked out why. First had come the discovery that there were parts of the course that I had never completely understood when I was at school myself. Revisiting them, knowing that I now had to master them, produced some eureka moments when explanations finally dawned on me. And these moments led in turn to spontaneous enthusiasm when I was in front of a class. After two years I had covered the complete syllabus and the magical moments of discovery had dried up: I concluded that I was not after all cut out to be a schoolteacher.

At just the right time, the autumn of 1968, when I had started my third year in teaching, I saw an advertisement for the position of 'science field editor' at the Clarendon Press, then the name of the academic publishing department of the Oxford University Press. The job briefly outlined struck an immediate chord. During one of my graduate student years I was the editor of the university chemical society magazine and I had loved every minute of it. A hankering to be in publishing must have formed and lodged itself in my subconscious, and then been awakened by the OUP advertisement. The essential qualifications asked for were the possession of a science degree and 'some publishing experience'. I sent with my application copies of the three issues of the university magazine I had edited, hoping that they would count as publishing experience. Apparently they did, and I eventually got the job.

The fact that the copies of the magazine seemed to have qualified as 'publishing experience', resulting in my getting an interview, may well, today, seem odd. In those days there were no publishing courses but, even so, the OUP then regarded it as normal to take on new graduates without experience and train them as assistant commissioning editors.

I joined the Clarendon Press in Oxford at the beginning of April 1969 as its science field editor. The job was to involve visiting scientists, chiefly in the UK, and chiefly in the universities, discovering what was going on, in teaching and in research, and finding new authors. Such a job couldn't possibly exist now—academic commissioning editors have long done their own campus visiting, and the main role of today's field reps is to promote textbooks—but it did then, at the OUP. It was a remarkable job and a remarkable time, not least because I enjoyed complete freedom, freedom to go where I liked and see whom I pleased: something else that couldn't possibly exist today.

And today this is something else that will seem distinctly odd. Why was I given this freedom? The first point to make is that there was then little commercial publishing experience at management level at the OUP. There was also a healthy submission rate of unsolicited proposals, and many proposals also coming from active academic series editors. The result was that there was little incentive (or indeed time) for busy editors to get out and find out for themselves what was going on in the academic world at large. Finding out was my job, as science field editor. And the reality was that my boss had no way of knowing what needed to be discovered: it wasn't possible to micromanage me, as we would now say, and direct me. That was the reason behind my remarkable freedom.

The Clarendon Press in those days was small, very small compared to what its successor, the OUP's Academic Division, eventually became and I found the atmosphere welcoming and friendly. It was fascinating during that initial period learning about how this complicated organization, the OUP, fitted together and operated. I thought it splendid that the three most senior people in the OUP's Oxford headquarters had official titles that did little to communicate to an outsider their importance: Secretary (in full, Secretary to the Delegates of Oxford University Press, the OUP's Chief Executive, Colin Roberts), Assistant Secretary (Dan Davin, head of the Clarendon Press), and Junior Assistant Secretary (Peter Spicer, head of the Press's schools publishing). Each fortnight during the university terms these three would attend the meetings of the Delegates, a board comprising the Vice-Chancellor, Proctors, Assessor, and at that time ten other Delegates, university dons. The Delegates considered each individual Clarendon Press book proposal and approved or rejected it for publication. One soon got used to various Press rituals, such as Morning Prayers. This did not involve any praying, at least not in my day, but was the name given to the examination and distribution of mail each morning. Mail would be opened by the office staff, the letters placed in a butler's tray on a stand, and Prayers would begin with the arrival of Dan Davin. Dan would chat to the assembled editorial colleagues whilst simultaneously scanning each letter, occasionally annotating one in his barely decipherable scrawl before putting it in an individual editor's pigeonhole.

There was time in those days for good, old-fashioned pedantry. I can remember receiving a memo addressed to a group of us—junior editorial colleagues—on the use of the semicolon. Copies of the letters we sent out each day were circulated within the group and some of us, it had been noticed, were failing to appreciate the full subtleties involved in the deployment of this particular punctuation mark. This is the OUP, the memo reminded us, and in such matters we should be 'beyond reproach'. But the most memorable example illustrating the temptation sometimes to consider taking things a little too far comes from a few years later. It was customary to have the Press Christmas card use an illustration from an OUP book published during that year, and inside would be a scholarly caption describing the picture. One year, in the mid-1970s, the cards had been printed and a tiny error in the illustration caption had been discovered. This was revealed at the academic division's weekly editorial meeting, chaired by Dan Davin. The error was so tiny and specialized that it would have needed, as it were, a medievalist with a magnifying glass to spot it. Nevertheless, the production department solemnly wondered at the meeting if an erratum slip should be printed to go out with each card. Sanity prevailed and the cards went out with the mistake, and no erratum slip. So far as I remember, no one wrote in to point out the error.

There was no training available on how to be a science field editor and during my first three months at the Press I did little else but read files: files on individual books or ideas for books, giving me an understanding of how initial ideas led to book proposals, some of which—a tiny minority, it seemed—led eventually to books. And during this time I discovered a series of files entitled 'Reports from Mr J. G. Crowther'.

Crowther's association with the OUP began in 1924 when he joined the Press as a representative for the technical department, newly created at the London office with the aim of exploiting the post-war boom in technical and vocational publishing. But by then that boom was over and Crowther, finding technical education in the UK to be in a depressed state, began to visit science departments in the universities. As Crowther himself put it in his memoir, *Fifty Years with Science*, published in 1970: 'The pure science departments of many universities were full of enterprise, hope and achievement. Scientists were prepared to write books, for which there was a considerable sale. In view of this situation, I did all I could to devote more and more of my time and energy to helping in the science publishing [of the OUP].' Whilst doing just that Crowther began in his spare time writing articles on science for the general public and, during 1927–1928, had some sixty pieces published in the *Manchester Guardian*. In 1928, he contacted the office of C. P. Scott, illustrious editor of the paper, seeking an audience, and was told that Scott might be able to spare him two minutes. Crowther records in his memoir the exchange that took place at the beginning of the interview. "Scott said, without asking me to sit down, 'Well, Mr Crowther, what can I do for you?' 'I wish to become a scientific journalist,' I said. Scott replied: 'The trouble is, Mr Crowther, there isn't such a profession.' To this I answered: 'I propose to invent it.'" The two minutes stretched into more than an hour and finally, a fortnight later, Scott wrote to Crowther to tell him they were happy to appoint him as their scientific correspondent. Crowther thus became the world's first scientific journalist. He also went on to write many books for the general reader on science and scientists. In 1930, his proposal that his engagement with the OUP should become a half-time appointment, to give him more time for his writing, was accepted, and his connection with the Press, eventually as a consultant, remained until well into the 1970s.

I read with growing fascination the filed reports from Crowther, describing his interviews with scientists in universities and research institutes during the 1950s and 1960s. The accounts were readable, never over-wordy, and full of insight. They provided context, they talked about the wider significance of the research being described, and the character of the scientist conducting it. That style, providing a genuinely interesting, rounded picture, left its stamp on me and this influenced how I approached my first job at the OUP back in 1969: talking to scientists and then writing reports for colleagues to read. I could still feel that influence right up to my retirement from publishing almost 35 years later.

I corresponded with Crowther during those first few months and became used to the formalities of old-fashioned salutations: Dear Crowther, Dear Rodgers. We used this style in conversation for many years. We finally met at the annual meeting of the British Association for the Advancement of Science, the British Ass, or simply BA, at Exeter in September 1969. I had never seen a photograph of him and had built up in my mind an image of the author of his reports and letters. The image wasn't even close. I met a genial, wiry, white-haired man of about 70. He had not just a sense of humour but a real sense of fun; his smile and his chuckle were infectious. He always wore the same clothes: dark blue suit, blue shirt, dark tie, black shoes (except at Flamborough, at the Crowthers' summer residence, when it was black trousers and navy blue fisherman's sweater).

I spent a lot of time with him that week. He showed me the ropes, such as getting me registered at the BA's Press Office. He advised me not to wear the Press lapel badge: one can find oneself in an unplanned conversation, he explained, and people will be on their guard if they see the badge; without it they are more likely to be open with you. And throughout the week there were the anecdotes: the people and the incidents from his life in the world of science. We met at every BA meeting throughout the 1970s and always there were the anecdotes. Over the years I listened to the same ones but that didn't matter; the way he told them, they were always enjoyable, and instructive. I remember being with him at one of the dinners at a BA meeting when he was so busy talking—an observation from someone sitting near would prompt a telling anecdote—that he still had a full plate when everyone else had finished. It was at one of these mid-1970s BA meetings that I remember his telling me of his astonishment that people were now actively seeking his views. The reason for this was that the political and social relations of science, and the notion of social responsibility in science (areas in which Crowther had been involved decades earlier) had re-emerged at that time. Crowther had become an elder statesman and this touchingly surprised him, and gave him a quiet satisfaction.

British Association meetings during that period were for me lively and exciting occasions and I looked forward to them as fertile sources of ideas for possible books, places to spot potential authors, and opportunities to hear accounts of ideas that could be genuinely thrilling. That said, there were plenty of deadly dull speakers and dire presentations to balance the occasional glorious moments, but no matter, the nuggets that popped up as unpredictable surprises made it all worthwhile. I think of one session and it can speak for them all. The 1978 meeting was held at Bath, and in front of me as I write this chapter is the programme for one morning's session, devoted to 'Fundamental physics'. The large lecture theatre was full, not so much to capacity as to bursting point. First, that year's president of the physics section, H. Elliot, Professor of Physics at Imperial College London, gave his presidential address on cosmic rays; then Martin Rees from Cambridge on galaxies and their nuclei; and finally J. C. Polkinghorne, also from Cambridge, on the quark structure of matter. It was wonderful cutting-edge science made memorable by first-rate speakers who were able in their different ways to communicate genuine intellectual excitement. I introduced myself to John Polkinghorne at the end of the session and asked if he could write as well as he could speak. He could and that marked the beginning of a collaboration leading to my commissioning his book, *The Particle Play: An Account of the Ultimate Constituents of Matter*, published in 1979.

Another kind of unpredictable bonus at BA meetings was having chance encounters with normally inaccessible people. The president of the physics section at the 1974 BA at Stirling was Professor Sir Hermann Bondi, Chief Scientific Adviser to the Ministry of Defence. His presidential address to the section was to be on 'Progress in gravitational physics'. Seeing this in the programme brought back memories of being in the audience some years before for a public lecture by Bondi, again on the subject of gravity. I clearly remembered Bondi's arresting style of delivery: he spoke with a pronounced Viennese accent, each sentence, uttered slowly, was perfectly constructed, the result being absolute clarity, and he spoke without notes. The lecture had been spellbinding from his first word to his last. Now, at the Stirling BA, circulating at the customary civic reception, following the formalities of the meeting's opening ceremony, I found myself joining a small group gathered around Bondi. During a brief lull in the conversation I caught Bondi's eye and here was an opportunity to ask a question. He had been Professor of Physics at King's College London since the mid-1950s and appointed a Government Chief Scientific Adviser in 1971. I asked him what differences he'd found working in Whitehall compared with academe. He looked at me through his large spectacles and I sensed that he was about to enjoy responding. He took a deep breath and in one second had collected his thoughts. 'It's like this,' he began. The next sentence had a theatrical quality. It came out even slower than usual, the gaps between each word slightly lengthened, and certain words were given heightened emphasis, these followed by even longer, dramatic pauses. 'When you have a . . . *disagreement* . . . you don't confront your . . . *enemy* . . . and leave with . . . *blood all over the carpet.*' For the next, concluding, sentence Bondi visibly relaxed, became more expansive, and he gazed into the middle distance in an almost dreamy way. 'You have a quiet word here . . . a quiet word there . . . and in a few months' time you find you have your own way.' Thus Hermann Bondi on Whitehall and its *realpolitik*.

The link between Crowther and BA meetings reached a high point at the 150th anniversary meeting at York in 1981 (the first BA, in 1831, was held at York, as was the 1881 meeting). Because of the importance of the anniversary, the BA had as its president that year a member of the royal family, the Duke of Kent. At the opening ceremony, held in York Minster, there was a degree congregation at which three honorary degrees were conferred: first on the Duke of Kent, then on Sir Andrew Huxley, the immediate past president, and finally on James Gerald Crowther. The applause for Crowther was much louder and longer than for the other two.

The Crowthers spent each winter at their flat in Bloomsbury and I sometimes visited them there. This is where I first met Francisca, Crowther's wife. One treasured memory from those times is of a meal at a restaurant. I had written to Crowther suggesting that the three of us might go out for dinner and his reply contained a splendid sentence that I've not forgotten. 'When we dine out in London we tend to patronize the modest establishment of Schmidt's, in Charlotte Street.' That is where we went and so I experienced just once the atmosphere of this 'modest establishment'. In my memory is a large first-floor dining room all but full of diners having animated conversations in a variety of European languages. The food was fairly basic but very good value. I mentioned the experience to my author Walter Gratzer many years later. Walter said he also had fond memories of Schmidt's, remembering a lady with a moustache at the cash desk, and a bad-tempered waiter who thundered down the stairs with heaped trays. On one occasion, recalled Walter, an aged waiter tripped and fell down the stairs in a cascade of china and sauerkraut and knocked himself out. The other waiters all stepped over him, or pushed him aside with their feet.

One quirk about the Crowthers was their choice of daily newspaper. Crowther was a Marxist—a gentle Marxist, as an OUP colleague of mine at the time described him—and yet the newspaper they had delivered was the politically conservative *Daily Telegraph*. This, they felt, was a perfectly sensible choice for them: it contained more news than any of the other papers and so it represented the best value.

The Crowthers had a cottage at Flamborough Head on the east Yorkshire coast. Here they spent every summer and I would stay overnight with them there when the annual BA meeting was in the north. At the end of the afternoon on those occasions the three of us would go into an adjoining field to gather mushrooms. I had excellent eyesight but Crowther was quicker than me at spotting where they were. Vegetables for that evening's dinner came from their garden.

Crowther was always interested to hear about the scientists I'd been seeing and what my general impressions were. On one occasion we were in the garden at Flamborough about to have a conversation when Francisca called to him to do something in the house. With a smile he indicated a bench: 'Sit here, Rodgers . . . and meditate.' On his return I told him that I was finding that while physicists and mathematicians were on the whole gentle and civilized individuals, chemists, especially organic chemists, could sometimes be brusque. A sage nod, then: 'Ah, you've noticed.' He'd discovered much the same thing decades earlier and had mentioned it to the educationist, Sir Percy Nunn. Nunn had immediately responded: 'Yes, yes, Crowther, it is obvious. Eminence in chemistry is an indication of a second-rate mind.' Crowther went on to explain to me that those were the days when the individuals who got a Nobel Prize for spending twenty years on a complicated synthesis of a natural product, say, were blessed more with dogged determination than with creative genius. That all changed when chemistry started dealing with the complex processes of life: these formidable challenges now attracted an intellectual elite. But at the time of our discussion in the garden, some chemists, like some earth scientists, did highly paid consultancy work for industry. This, Crowther suggested, explained the behaviour patterns I'd noticed.

An example of Crowther's mischievous but gentle sense of humour comes to me from those times. I'd driven him back to Flamborough after a BA meeting and we were having a cup of tea before my departure for home. During Crowther's absence that week Francisca's sister had been staying. An animated Francisca declared: 'We were talking until two in the morning.' Crowther gave me a good-humoured smile and then turned to Francisca and asked: 'But my dear, were either of you listening?'

A final, vivid memory of those Flamborough Head days: crabbing. I'd taken my family to spend a day with the Crowthers. It was late-May or the beginning of June in 1973, when my wife was more than eight months pregnant. The way down to the beach was by means of a rope ladder down part of a cliff. Sticking in my memory is a picture of a heavily pregnant wife, two young sons, Crowther in his 70s, and Francisca climbing in turn down a cliff by means of a rope ladder. Crowther was an expert crabber, using a special wire implement he had constructed for coaxing out crabs he had seen disappearing into small crevices in the rocks. We eventually climbed back up the rope ladder with a splendid catch, which Francisca started to cook as soon as we were back at the cottage.

At some stage in the late-1970s our formal mode of address, 'Crowther' and 'Rodgers', changed to 'Jim' and 'Michael'. We continued to correspond after Crowther finally retired from his association with the OUP, and the last letter I received from him was in December, 1982. In it he remarked on the fact that he and Francisca were feeling the effects of aging more and more, as the time passed. 'Still,' he concluded positively, 'on the whole, I get along not too badly.' He died not long after, and desperately sad news followed. A distraught Francisca, unable to contemplate life without him, threw herself off the cliffs at Flamborough Head.

Crowther's association with the Oxford University Press lasted for some fifty years. How significant was his contribution to the science publishing of the OUP? At the start of the 20th century, the OUP's publishing was centred predominantly on the classics, humanities, and bibles, though science was by no means entirely absent: the OUP's first publication in modern science, in 1873, was no less than the classic *Treatise on Electricity and Magnetism* by James Clerk Maxwell, the pre-eminent genius of 19th-century physics. At the time of Crowther's appointment in 1924 as a representative for the London office's technical department, and until after the Second World War, no one with scientific training was employed by the OUP, but senior managers at the Press began to listen to his recommendations. In 1927, for example, Crowther's advocacy was largely responsible for the OUP hearing about and subsequently publishing a translation from Russian of Pavlov's seminal account of his work on dogs salivating to the sound of a bell in the absence of food, having previously been conditioned to associate the sound of a bell with the arrival of food, *Conditioned Reflexes: An Investigation of the Physiological Activity of the Cerebral Cortex*. In the previous year, at the Oxford meeting of the British Association, Crowther was in the audience for a lecture by Arthur Eddington on 'Stars and Atoms'. Afterwards, Crowther called on Eddington and persuaded him to combine the lecture with some others he had given and have them published by the OUP. The result, *Stars and Atoms*, published in 1927, the first of the famous Eddington and Jeans bestsellers, was the Press's first 'popular science' book. Crowther attempted but failed to persuade Jeans to publish with Oxford. In his 1970 memoir, *Fifty Years with Science*, Crowther recounts a meeting he had with the illustrious Ernest Rutherford, founder of nuclear physics, some years after the publication of *Stars and Atoms* (Crowther enjoyed a good relationship with the great man), when Rutherford told him: 'Jeans said to me "that fellow Eddington's written a book which has sold 50,000 copies. I will write one that will sell 100,000", and by God he did.' The Eddington and Jeans bestsellers were widely read popular science books written by two English scientists, the astronomer Arthur Eddington

(1882–1944) and the mathematician and astrophysicist James Jeans (1877–1946).

Crowther's greatest and most enduring achievement for the OUP was his crucial part in the creation of the International Series of Monographs on Physics. Its origins lie with a conversation Crowther had at the 1926 Oxford meeting of the British Association with the Oxford chemist and the OUP Delegate, N. V. Sidgwick. Crowther asked him to suggest possible authors and Sidgwick recommended P. Kapitza, a Russian physicist working on magnetism at the Cavendish Laboratory, the physics department at Cambridge University. On visiting him, Crowther discovered that Kapitza's real interest, rather than a book on magnetism, was in a series of advanced books on physics, to be planned and edited by himself and R. H. Fowler. The result was the OUP's International Series of Monographs on Physics, edited by Kapitza and Fowler. Fowler persuaded his former pupil P. A. M. Dirac (1902–1984) to write for the series, and the outcome was the classic *The Principles of Quantum Mechanics*, the first title to be published in the series, in 1930. Graham Farmelo's magnificent biography, *The Strangest Man: The Hidden Life of Paul Dirac, Quantum Genius*, published in 2009, brought home to a general audience just how remarkable Dirac was. A pioneer of quantum mechanics and winner of the Nobel Prize for Physics (in 1933), he is acknowledged as the greatest British physicist since Newton. Einstein regarded him as his equal.

Crowther's memoir provides a charming picture of Dirac shortly after his book had been signed up by the OUP: 'When I first called on Dirac he was living in a simply furnished attic in St John's College [Cambridge]. He had a wooden desk of the kind which is used in schools. He was seated at this, apparently writing the great work straight off.' Remarkably, Dirac's book is still in print. At the time of writing, the current, fourth edition was published in hardback in 1963. A paperback edition came out in 1981 and sales of this to date exceed 25,000 copies, with an annual sale of around 500 copies. The International Series of Monographs on Physics remains to this day the flagship of the OUP's physics list. At the beginning of 2013, title number 158 was published, with further titles waiting in the pipeline.

Crowther had no interest in the business side of publishing, which is doubtless why he had no wish to become a book publisher's editor. What he wanted to do was visit scientists, talk to them about their work, and identify and attempt to persuade those with a good book in them. The enlightened senior managers at the OUP, not one of them with any scientific qualifications but recognizing Crowther's value, had the good sense to give him the freedom to pursue his interests. They read his reports and, with the science Delegates, were prepared to take seriously his recommendations. After the Second World War, the OUP appointed a science editor and as the number of editorial staff with science qualifications gradually increased, the relative importance of Crowther's contributions decreased. But the impressive foundations for which he was largely responsible were in place. Crowther's legacy was substantial, and it is today still yielding dividends.

The high rates of inflation during the 1970s led to a rapid escalation in the price of books, and much else too, of course, and this was debated briefly in the correspondence columns of *The Times* in May 1972. It began with a letter from the chairman of the Brighton Poetry Society, expressing alarm at the soaring price of books. Books, he wrote, 'must be far beyond the purchasing power of the average wage earner in this country, including students, young intellectuals and the general population'. This was seized on by a critic of the OUP, who pointed out—rightly—that scientific books were affected perhaps worst of all by the problem of price inflation, but went on to claim that books produced by the Oxford University Press 'have suffered more than [those from] any other [publishing house]'. In evidence, he cited two books in the Press's International Series of Monographs on Physics. A standard work on the theory of relativity cost £2.75 in 1970, he wrote, and a new edition of the same work 'has recently appeared containing a few more pages and costing £13'. Worse, an important new monograph on solid state physics had just been published 'at the astronomical price of £28'. The letter concluded thus: 'Equivalent books can be purchased at railway bookstalls in the Soviet Union for 50p! Presumably scientific books are heavily subsidized by the Soviet authorities, and it would be gratifying to see the same sense of priority shown in this country.' This drew a response from Colin Roberts, Chief Executive of the OUP. He called into question the contention that the prices of books published by the OUP were higher than those of comparable works published by other houses. There was no evidence to support this, he said, though if it appeared to be so, this might be because the Press had deferred putting up prices in some areas longer than some other firms. But he agreed that book prices were 'deplorably high' and that the high rate of inflation had a particularly damaging effect on science books, with their demand for complex typesetting and illustrations. He dealt briefly and persuasively with the specific criticisms relating to the two physics titles. The gap between the first and second editions of the relativity book was not two years but twenty; and the price of the solid state physics book was perhaps understandable given that it contained more than 750 large-format pages, with a large number of complex tables. And then a lovely riposte, and the sole reason why I can remember this exchange. Roberts ended his letter: 'If it is really the case that a comparable Russian book is sold on railway bookstalls for 50p, it is a nice question whether the lavishness of Soviet subsidies or the stamina of their railway passengers is more to be admired.'

The problems caused by the high rates of inflation in the 1970s bring to mind the bad times endured by the OUP at that time. The message from the accountants was exceedingly simple: cash coming in from the sale of books in today's money was insufficient to pay for the printing of new books at tomorrow's inflated costs. Manuscripts leaving editors' desks for the production department had to be rationed because funds for manufacturing new stock were strictly limited. In addition, editors were asked to negotiate the termination of authors' contracts when the planned delivery was overdue. It was, in short, a depressing time. I felt that my best years might be frittered away, my time spent applying brakes rather than going out in search of new authors. But the Press survived and after a low point in the mid-1970s things steadily improved.

In addition to Crowther and the BA meetings, one other thing I associate with those early years at the Press is being sent on management courses. Decades later it finally dawned on me why they never really worked for me—I wasn't blessed with the right sort of management thinking apparatus—but I do have fond memories of them. In the autumn of 1972 I journeyed to a management centre on the outskirts of Bromley in Kent to spend the best part of a week studying 'Progress to Profit', as the course title had it. One striking fact about the mix of delegates attending became apparent once the routine of the opening session with its introductions was under way. The great majority, most from small or medium-sized publishing houses, were managing directors or editorial directors. The Oxford University Press, with its flat pyramid of a management structure, had despatched two of its very junior members, an editor from the music department at the Press's London office, Anthony Mulgan, whom I had not met before, and myself, the science field editor.

There are two incidents I can remember from the course. The first, of no consequence but it still makes me smile, was early in the week when a small group of us spent an evening after dinner making use of the centre's bar, which had a snooker table. Christopher MacLehose, then editorial director at Barrie and Jenkins, was there and he impressed me with his panache on the green baize. We retired late and in consequence just missed breakfast in the centre's refectory. One of our number came from Bromley and assured us that it was but a short walk to a nearby transport café. A busy greasy spoon, it was presided over by a middle-aged lady who, with a spatula, was attending to the contents of various frying pans. In my mind she has a cigarette hanging from her lips but that is doubtless my memory playing tricks. Christopher MacLehose was the first in our queue and he asked in his carrying fruity voice: 'Do you have cornflakes?' The establishment seemed unused to hearing such rich, patrician tones and the volume of conversation instantly dropped from *forte* to *piano*. The lady behind the counter, startled by the unusual request, and the manner of its delivery, was rendered speechless and she simply stared at Christopher. I cannot now recall the outcome but I think Christopher had to make do with a fried egg or bacon sandwich like the rest of us. Difficult to imagine any of this in our modern world! (Decades later, Christopher set up MacLehose Press, under the aegis of the Quercus imprint, devoted to the translation of literature and crime fiction into English. Shortly after, a Stockholm publisher approached him with two translations in typescript, with the promise of a third soon to follow. The first of this trilogy, published by MacLehose Press in 2008, was *The Girl with the Dragon Tattoo!*)

The other incident, which does have relevance to management methods, took place on the last morning of the course. This session was devoted to 'leadership' and the organizer described the exercise on which we were about to embark. It was to involve groups of us building towers with interlocking plastic blocks of Lego. ('Oh, God,' cried out my colleague Anthony Mulgan, memories of his national service days plainly flooding back. 'This is what they made us do in the Army.')

I need to introduce two fellow delegates at this point. One was Sidney J. Josephs, head of Macmillan Press in Basingstoke, known by everyone as Joe Josephs. I had never met him before the introductions at the beginning of the week but had heard stories about his authoritarian style and the fact that he didn't suffer fools gladly. It was easy to see as the week progressed, observing his provocative style in discussions, that his staff might well find his manner slightly intimidating. The other delegate who forms part of this story was from Penguin. In charge of their reprints, he was a pleasant, exceptionally quiet individual.

We were divided into small groups, each given a separate room containing a large supply of Lego, and asked to begin by choosing a leader and an observer. The object was to erect a tower against the clock, measuring its height when time had run out. The leader began by encouraging a discussion on the strategy to be adopted and then we built the tower, all this under the gaze of our observer. Time was called, our tower measured, and we all trooped back into the main lecture room. In turn each observer gave an account of what had happened, the style adopted by the leader, the cooperation dynamics of the group, and so on. It was all highly unremarkable until one particular group reported. The taciturn man from Penguin was its observer, Joe

Josephs its leader. An absolute silence descended as we listened to the quiet account from the observer. He described in clinical detail how the leader's abrasive style had led to a gradual disintegration of the group, the climax being the collapse of the tower. Joe took it on the chin, listening impassively. We all averted our gaze, wondering perhaps if a Lesson in Life had been learned. Looking back, it was certainly unusual to have witnessed the effects of bullying under, as it were, laboratory conditions.

Some twelve years later I related this story to Donald McFarland, science editor at Penguin. Donald was about to move on from Penguin and we were having lunch to mark the event. I learned that the Penguin man in the story, the observer, had progressed to become head of Penguin's warehouse and distribution centre. Donald then described a routine management meeting at Penguin, chaired by its renowned head, Peter Mayer. The erstwhile observer treated the meeting to a catalogue of frightful problems besetting the distribution operation. When he'd finished, Mayer signalled his empathy with a sympathetic look, put a hand on his shoulder, uttered the words, 'Sort it out for me, John', then moved immediately to next business on the agenda.

Very many years after that, towards the end of my time in publishing, my assistant Abbie Headon would occasionally apprise me of some complicated problem, often requiring for its solution IT skills wholly beyond me. I would put my hand on her shoulder and say the magic words: 'Sort it out for me, John.'

CHAPTER THREE

FALLING UNDER THE SPELL
OF THE
SELFISH GENE

IN THE SUMMER OF 1972 management consultants, a firm called Cooper Brothers, descended on the Clarendon Press, their mission to identify what might be improved, from the minutiae of systems and procedures to grander matters of publishing policy. Two consequences, one from each of those extremes of that spectrum, have remained in my memory. The detailed one concerned the Press's filing system, a subject normally guaranteed to lead instantly to the glazing-over of eyes. It actually led to the end of something rather wonderful, and it produced too a memorably comic moment.

All files at that time were located in a central general office and the filing system had the sort of complexity that can be created only by decades, if not a century or two, of evolution. Whenever an author wrote to the Press with a proposal for a book, or an editor wrote to a potential author with a suggestion for a possible book, a Preliminary Proposal, a PP file was started. There were very many such files, and the number identifying a new PP file had reached five digits during this period. The majority of these proposal ideas came to nothing, their files languishing for untold years in the PP section, but a select few progressed to the next stage, formal proposals ready for a publishing decision at a meeting of the OUP's Delegates. These became Proposals, the paperwork transferred to a P file and moved to another part of the office. Formal acceptance by the Delegates ('encouraged' was, and is the term used) meant an advance in status to a Book in Preparation, with a new BP file in a different location. Many BP files were destined never to progress any further, their authors failing for a variety of reasons to finish the job. But a number of BPs were eventually completed and delivered, and this meant a final change of status in the filing system. It was at this stage that a book, in the form of a typescript ready for editing before being sent off for typesetting, was allocated its Standard Book Number, the SBN. The 9-digit SBN code was invented in the UK in the mid-1960s. This pioneering version became the 10-digit ISBN, developed by the International Organization for Standardization and published as an international standard in 1970, though the 9-digit SBN code was used in the UK until 1974. And to round off this particular story, ISBNs have comprised 13 digits since the beginning of 2007.

That is quite enough about ISBNs and I now return to the Cooper Brothers consultants at the Clarendon Press in 1972. The Press's elaborate filing system was an easy target for them and they proposed simplifying the brilliantly logical sequence, with its steady progress from PP to P, then to BP, and finally to SBN, recommending the retention of just two of the stages, P and SBN. These and other recommendations were presented at a meeting attended by the editors. One of our number mourned the passing of PP files, explaining that he, like other commissioning editors, routinely wrote letters to many potential authors suggesting ideas for possible books. A PP file was the ideal repository for the office copy of a letter of this kind; where would the file copies of such letters now go? Ah, said the consultant, and we watched as he pondered the tricky, unanticipated question. Inspiration came after no more than two seconds. The solution is simple, he declared. File them under B for Book. After the resulting hilarity had eventually died down, an editorial colleague recalled an incident from an earlier career when he had worked in a government department in Whitehall. A typist inexperienced in office procedures had been brought in as temporary help, and some months after her departure the file copy of a letter she had typed was urgently required. The office staff looked in every conceivable place for it but all such attempts failed until eventually someone had a brilliant idea. There it was, filed under L for Letter.

The other remembered part of the Cooper Brothers exercise was to have for me far-reaching consequences. Editors were asked to suggest how the Press's publishing might be improved or extended and each publishing department was invited to submit a paper setting down these ideas. The science department of the Clarendon Press then comprised a science editor, three assistant commissioning editors, and the field editor, me. The most recent to have been recruited was Adam Hart-Davis, who had joined the OUP a year earlier, in 1971, as college (meaning textbook) editor. Adam was given the job of coordinating the science department's response: we all wrote down our comments and ideas and he combined everything in one paper. One of my suggestions was that the Press should consider embarking on a popular science publishing programme. I am not able to pinpoint anything specific that had ignited my interest in popular science: it was more a gradual process unfolding over the previous year or so. I can remember being gripped by a few lectures I attended at British Association for the Advancement of Science annual meetings, and early in 1972 being intrigued by a memo from Dan Davin, head of the Clarendon Press, inviting comments on a possible Oxford Companion to Science. So the timing of the Cooper Brothers request for suggestions was right and I enjoyed setting down the reasons why I thought popular science publishing would be good for the Press.

The idea, popular science publishing, came up again a few months later at a meeting of the science editors with the science Delegates, in December 1972. It was welcomed as something worth looking into and all agreed that I should start the ball rolling by preparing a paper on popular science publishing. The resulting piece, completed in April 1973, was circulated for comment to Colin Roberts, Chief Executive of the OUP, and his deputy, Dan Davin, the science editors at the Clarendon Press, and the sales and marketing departments, then housed at the OUP's London office. The opening paragraph gives a taste of a time before the explosion in interest in popular science book publishing that was to come:

> During a visit to the United States just over a year ago I talked to two
> scientists who both, without prompting, made the point that unless the

general public is given a better and more balanced understanding of science then there was a real danger that research funds would eventually dry up. One of them, a physicist, said that ten years ago scientists could say: 'We want to study mesons because they are interesting' and that was sufficient to get them the money. Today, however, this is not enough and people want to know why it is that mesons are interesting and why they are worth spending money on. I remember wondering at the time if this might mean that distinguished scientists, who at the present time generally leave the writing of popular science to professional writers, would begin to feel that writing for the non-scientific public was something they ought to start thinking very seriously about.

I made it clear that I was not attempting to set out a cut-and-dried plan ready for immediate implementation. 'There are enormous problems,' I wrote, 'and I must confess that I am not yet sure that what I am suggesting can in fact be done. What I have done is to put down some ideas and indicated some of the unsolved problems. I have also provided notes on some relevant interviews.' What I proposed at the end of the note was a series of books in various fields of science: these were general areas, not specific topics.

The initiative was to have one specific, very special consequence, though not until three years later, and this was to change my life. My name had become linked with an enthusiasm for popular science and this was why a letter from Roger Elliott, a physicist and a science Delegate (and later to be Chief Executive of the OUP), landed on my desk on 23 February 1976. Handwritten on New College letterhead, it was short and to the point:

Dear Michael

One of the dons here, Dr C. R. Dawkins, is writing a popular science book tentatively called 'The Selfish Gene' which he describes as in the genre of the 'Naked Ape'. I have no idea whether he or it is any good but it might be worth looking into. I got the impression that he is looking for a publisher.

Yours ever
Roger

I tried ringing Dawkins at the zoology department several times but without success and so wrote to him later that week asking if I might see the material so far written. In the middle of the following week, 3 March, I received a memo from Dan Davin reporting on a conversation he had had the previous day with the philosopher Anthony Quinton, an OUP Delegate and a fellow of New College. Quinton had told Dan that 'a New College man' was writing 'a rather lively book on the function of the gene . . .'. I wrote a response on Dan's memo, sending this back to him, explaining that I had already heard about it from Roger Elliott, had written to Dawkins and was awaiting a response. My message ended: 'All I have at the moment is the title, which is quite lovely: THE SELFISH GENE.'

Dawkins's response to my letter came at the end of the next day: a brief telephone call from him to let me know that a package containing eight of the eleven planned chapters would be dropped off at the OUP's porter's lodge in time for me to collect on my way home. Ominously, Dawkins's accompanying letter told me that Tom Maschler of Jonathan Cape had seen most of what had so far been written. But—a ray of hope—Dawkins explained that he would be interested in discussing things with a 'slightly different' kind of publisher.

I knew before reaching the bottom of the first page of the opening chapter that here was something quite extraordinary, and I can remember the sharp jolt as the expectation of perhaps no more than an interesting read in prospect suddenly changed: the writing had reached out and grabbed me by the lapels. I can remember, too, periodically taking my eyes off the typescript, not simply to let the ideas sink in but also to allow the nervous energy of excitement to dissipate. This might all sound like a ludicrous over-reaction but it is how I remember the experience. Those who over the years have enjoyed reading *The Selfish Gene* will know why it works so well, why it casts such a potent spell: the writing, so fluent and with its wonderful rhythms; the big, intellectually exciting ideas, rich and complicated but so beautifully explained; and throughout, the marvellous stories from animal behaviour regularly brought in to illuminate points along the way ('alive with fascinating stories', as we put it in the eventual jacket blurb: ' . . . about fish who queue up to have their teeth cleaned—and then refrain from swallowing the tiny dentist; about ants who take slaves and tend fungus gardens; about the kamikaze bees who commit certain suicide when they sting robbers of the communal honey'.) It was all completely new to me: I was trained in the physical sciences, my degrees are in chemistry, I had never been taught any biology, even at school, and this opening up of a wholly different world heightened the intensity of the experience. Something else that seemed very clear to me as I read the opening pages was that here was a book that was going to create a stir. By the time I had finished I could say that, yes, reading the chapters had been exhilarating, and that yes, the whole thing had taken a powerful hold over my imagination. But, as an editor, what was really intoxicating was feeling wholly convinced that the book was going to make waves. It was going to sell.

I arrived at the office the following Monday morning, 8 March, full of enthusiasm for the book, and at the same time also sick with worry that I might lose it, to Cape or another high-profile publishing house in London. The combination of excitement and worry had been keeping me awake at night but the enthusiasm easily won and supplied me with all of the energy I needed. I telephoned Richard Dawkins and told him I thought the chapters were terrific and that I had loved reading them. We arranged to have lunch at the Press later in the week, on Thursday. Then I rang Professor Jim Gowans.

Jim, a distinguished Oxford immunologist, was the OUP Delegate with special responsibility for biology. His support would be essential if I managed to get the book as far as a formal proposal and put to the Delegates for their approval. I told Jim about the book and the effect that reading the chapters had had on me. If I got the chapters to him today, would he read as much as possible and get back to me with his reaction as soon as possible? He would.

At the end of that afternoon, wholly fortuitously, a meeting had been arranged with David McFarland to talk about a book he was working on for the Press. David was a senior academic in the zoology department and so a colleague of Richard Dawkins. David knew about the book though hadn't actually seen any of it. He confirmed that Dawkins's scientific judgement was sound. The views he was putting across, David went on, represented the picture as seen by modern biologists today. So, Dawkins was not an eccentric out on a limb? No, confirmed David, he was simply a good expositor wanting to gain access to a wide lay audience. David had a final snippet of information and it greatly cheered me. Another publisher, he said, and I took this to be a reference to Jonathan Cape, thought that the present version was too intellectual for a mass readership.

Jim Gowans rang me the next day. He had been able to read about half of the eight chapters and liked them a lot. That said, it was not his field and so he said he wasn't able to give the book a clean bill of health from the academic point of view. If there *is* a scientific flaw, Jim said, it would need someone like the geneticist Walter Bodmer to spot it.

I met Richard Dawkins for the first time on Thursday, 11 March, and over lunch I did my best to convince him that the OUP could rise to the occasion and do justice to the book. My proposal, I said, would be to put the estimated extent of the final version of 65,000-odd words into 200 pages, publish in the following spring in hardback with a selling price of £2.95, cheap enough to encourage the maximum sale, and with a paperback in perhaps 18 months' time.

We talked about Cape. Richard's fear about going with them was that they might want to over-sensationalize the book. One appeal of the OUP, he said, was that its imprint would confer a stamp of respectability. Also, Tom Maschler didn't like the proposed title, thinking that 'selfish' was an off-putting word. I said I thought that juxtaposing it with 'gene' was so unexpected that the resulting title was positively arresting. Meanwhile, Desmond Morris had suggested *The Gene Machine* as the book's title. 'That is a more saleable title than *The Selfish Gene*', Jon Stallworthy jotted down on the memo I subsequently wrote summarizing this meeting for Dan Davin and his deputy Jon, my immediate boss. I ended my memo: 'Well, Dawkins is now going to talk to his wife, to Desmond Morris, and to David McFarland (I've alerted him to do what he can). Also, the bit I don't like, he is going to let Tom Maschler know about the present situation. I can just see some people doing the OUP down, saying we're not cut out to handle a book like this. So I tried to convince Dawkins about the changed attitudes here and so on—preparing him for possible comments based on our past reputation.'

That evening I rang John Lord at his home. John was the trade marketing manager based at the OUP's London office, Ely House in Dover Street. He asked me to get a copy of the chapters to him the next day, promising to read them over the weekend. And he agreed to come over to Oxford the following week to meet Dawkins and help in trying to persuade him to come to us.

I contacted Richard the following morning and asked him if he would postpone making any decision, and not talk to Tom Maschler, until after a meeting I wanted to arrange with John Lord the following week. Richard agreed. He went on to say that he had been thinking carefully about the possibility of rapid publication, meaning the coming autumn. He said he was afraid that the subject might well take on bandwagon proportions and that next spring could be too late to make the maximum impact. In principle this could be managed and I said we would discuss it when John Lord joined us the following week.

At John Lord's suggestion I had a copy of the first few pages of the typescript despatched by express mail to the OUP in New York and then phoned my colleague there, Bill Halpin, to alert him. If New York like the sound of the book, I said, and let us know this at the beginning of next week, we could report it to Dawkins when John Lord and I planned to meet him for lunch. This could help to sway his decision. Bill agreed to read the pages as soon as they arrived and said he would hope to telephone me on Monday. He did but alas was cautious: was it 'just another book?' It was agreed that I would send over the eight draft chapters so that he and his colleagues could get a proper feeling for the book. In my covering note sent with the chapters I wrote: 'Forget your worries about this being just another book, as discussed on the phone. This is new ground as far as the general reader is concerned. In fact if you don't sit up half the night finishing off the typescript and then concluding we have something really terrific on our hands . . .'

John Lord and I had lunch with Richard on Wednesday, 17 March. It turned out that John, like Richard, had read zoology at Oxford, a couple of years ahead of him. I took this to be a good omen but something even more extraordinary was to follow. Richard had travelled to Oxford as a schoolboy to take the entrance examination and it was discovered that he had come without a set of laboratory instruments needed for the practical. An undergraduate in the college who could lend Richard the required equipment was found, and that undergraduate was John Lord. I doubt that this odd coincidence played any serious part in helping to make up Richard's mind to sign with the OUP, which he agreed to do over our lunch, but it is nevertheless a pleasing tale.

The following morning I travelled to Long Crendon, a village about a dozen miles from Oxford, to visit Norman Gowar at his home there. Norman was a lecturer in mathematics at the Open University and I had commissioned him to write a book on mathematics for the general reader. As the opening of the eventual jacket blurb had it, 'Many people are frightened of mathematics and yet would like to have some understanding of what the subject is about. This book is for them.' And Norman was the man to write it: *An Invitation to Mathematics*, published in 1979. The routine Norman and I had established involved his sending me a draft chapter which I would edit, meaning identifying exactly where I started to find the going tough, or where I simply didn't understand something, or felt in need of an example to push home a point, or wanted a slowing down, or a pause with a 'taking stock' summary, and so on. We would arrange a date for me to come over to Long Crendon, go over the edited chapter or chapters, agree on the changes Norman would make in the revised version, and then walk down to the local pub for a beer and sandwich lunch. As soon as we got to the pub on this particular day, our working session completed, I began to tell Norman about the Dawkins book. I enjoyed having him as a captive audience, and he was a good listener. Eventually confessing himself intrigued he casually said that he'd mention it to Vivienne King, someone he knew at the BBC. Norman regularly gave lectures for the Open University, filmed and transmitted by the BBC, and so he had a number of contacts there. I didn't really register the significance of this, returned to Oxford, and thought no more about it.

Just over a week later, on Friday, 26 March, I was telephoned out of the blue by Vivienne King. She said she was calling from the BBC's unit responsible for making the *Horizon* science documentary films, and that she had heard some details of the Richard Dawkins book from Norman Gowar. Could I tell her a bit more about it? Vivienne went on to say that they had just begun to think about a *Horizon* programme in the genetics area for transmission in the autumn. I promised to put in the mail a copy of a blurb we were in the process of polishing in order to give her a clearer idea about the book. Vivienne was on the telephone to me again on Monday morning. The blurb had arrived, they were all fascinated, and so on. Then a request: she and Peter Jones, a producer for *Horizon*, would like to come to Oxford for a discussion and also meet Dawkins. Of course, I said. When did she have in mind? Her response came back immediately. 'How about tomorrow?'

Tomorrow it was and I took the three of them for lunch at the Trout in Godstow, just outside Oxford. Before the arrival of the BBC duo Richard repeated a point he had made before, that he was not the originator of the basic ideas described in the book. For this reason, he told me, he wished to take something of a back seat in any television presentation. I think too that he felt some nervousness at the prospect of being in front of the camera for the first time: would he be able to carry it off, or would he freeze? Peter and Vivienne asked early on over our lunch if he would consider presenting the programme and, on receiving a tentative no, immediately switched to asking who might be suitable. The answer to that was John Maynard Smith, JMS. Richard later wondered if he'd done the right thing but there was no turning back: Peter and Vivienne now had JMS firmly in their sights. The best we could hope for was Richard eventually being listed in the credits as the author of the book, *The Selfish Gene*, and most important of all, that the film would actually be called 'The Selfish Gene'.

Peter and Vivienne said they now had to persuade *Horizon*'s editor, Michael Goodliffe, but they were so completely sold on the idea that this seemed to be a formality. They took away a couple of copies of the typescript and said they would now outline a possible approach for the programme for discussion with Richard.

Formal acceptance of the book for publication required academic reports confirming that the science was sound. I rang Professor Walter Bodmer, head of Oxford's genetics department, and asked if he would read the Dawkins typescript in a hurry to check for scientific flaws. Walter agreed to do what he could and said that in addition a copy ought to be read by an American population geneticist working in his department, Dr Gleynis Thomson.

Walter telephoned me on Friday, 26 March. He said that he and Gleynis Thomson had picked up a number of points they thought were wrong and that the best way of dealing with this was for Richard to have a session with them. Walter wanted in the meantime to give me an instant reaction. His central point was that the simple parts of the book would be too simplistic to satisfy a professional audience, and that the passages dealing with the rather sophisticated concepts that Dawkins covered would be too sophisticated for a lay audience. In short, there was a danger that it would fall between two stools. I set down a summary of this conversation with Walter in a note for Dan Davin and Jon Stallworthy, and on this particular criticism of Walter's wrote: 'This is a question of publishing judgement and although it is a fair point I do not myself agree with it. My view is that the book is so well written and so compulsively readable that whenever sophisticated concepts are dealt with there is a positive momentum carrying the general reader along. In other words, he has a positive incentive to follow through the arguments, which are put over in a very readable fashion.'

On the following Monday morning, 29 March, I went round to the genetics department and spent an hour or so with Gleynis Thomson going through her annotated copy (and also Walter's copy) of the typescript. Walter came by for a brief word and it was arranged that Richard should have a working session with them in the department that afternoon. In my note on my meeting with Gleynis I wrote: 'The central criticism seemed to be this. Dawkins argues that one can talk about individual genes. These are the potentially immortal replicating units the book is all about. Gleynis Thomson and Bodmer say that one cannot really talk about individual genes: one must talk about all the genes in an individual body. One cannot separate out the operations of individual genes.'

I asked Richard to give me a ring when his afternoon meeting with Gleynis and Walter had finished. He gave me an account of how it had gone over a drink early that evening in the Royal Oak in Woodstock Road. He had spent three hours with them and they had gone over each of their points of criticism. Richard told me that virtually all of these points stemmed from the one central issue about whether one talks of individual genes or the so-called genome—all of the genes in a body taken together. 'This,' I wrote in my note on the meeting, 'is a point of controversy. After talking to Dawkins I am convinced that this is what it is rather than a black-and-white right or wrong issue.'

My note continued: 'If Dawkins took the Bodmer–Thomson view to its logical conclusion it would completely emasculate the whole message of his book. Bodmer and Thomson raised a whole host of points of detail and Dawkins will incorporate their suggestions in his revised version. But he will stick to his view that one can talk about individual genes. He will make it clear at the beginning of the book that this particular point of view is not a universally accepted one.' In fact, Richard had a subsequent meeting with Gleynis and he sent me a detailed letter setting down exactly how the disagreement, semantic rather than substantive, had been resolved. An early chapter, the main source of the problem, had now been substantially rewritten, with Walter's views very much in mind, making Richard's meaning completely clear, and removing the misunderstandings that had arisen from the first version. There were now no longer any grounds for doubting the book's academic credentials and the OUP's Delegates accepted it for publication at their first meeting of term after the vacation, on 27 April.

Meanwhile it was agreed with John Lord that we should publish the book in the autumn as a 'star title'. A production schedule had been worked out and a provisional publication date was set for 14 October.

There was much agonizing over the book's title. I loved the original *The Selfish Gene* from the moment I first read it in Roger Elliott's letter.

For me it had a brooding presence and I sometimes imagined the words suddenly appearing on a screen, as if the title of a film, accompanied at the same instant by the doom-laden opening chord of Mozart's overture to *Don Giovanni*, if that doesn't sound too fanciful. But the trouble with having the word gene in the singular, argued some colleagues, is that it implies one mutant, rogue gene amongst a population of normal ones. At an early stage John Lord suggested *Our Selfish Genes* but, over our drink in the Royal Oak, Richard rejected this, though said he would settle on the compromise *The Selfish Genes*. Other colleagues felt strongly that we should go for Desmond Morris's suggestion, *The Gene Machine*. I argued against this, ending my note describing the sessions involving Gleynis, Walter, and Richard: 'I can see clearly all the advantages but it is simply the wrong title. It does not convey the central message of the book that genes behave as if they were selfish. *The Gene Machine* is neutral.'

There was much agonizing too over the jacket, and we struggled with a variety of approaches, from images of birds with huge gapes to a scene of aggression in a children's playground, but nothing seemed to work. Then Richard telephoned me out of the blue to tell me that he had just been talking to Desmond Morris. In his spare time, Richard began, Desmond was a Surrealist painter, his paintings influenced by biological shapes and structures, and Desmond wondered if we would be interested in looking at the ones on display around his house in Oxford, and perhaps choosing one for the jacket of *The Selfish Gene*. Richard and I were joined by Andrew Thomson, the OUP designer to be responsible for the design of the book and jacket, and we walked over to Desmond's house at the end of an afternoon at the beginning of April. His paintings were on display all over the house and I think our tour took in every single room. Eventually we converged on one painting, with everyone agreeing that this was the one—its strong central focus would clearly lend the finished design a powerful impact (see Plate 1).

Without further ado Desmond transferred the chosen work to an easel, strategically placed underneath a spotlight, and then produced gin and tonics for us to sip whilst we gazed at 'The Expectant Valley'.

In 2006, the OUP published a special 30th anniversary edition of *The Selfish Gene*, pleasingly reproducing this original jacket illustration.

Walter Bodmer and Gleynis Thomson read Richard's first draft in order to identify problem areas with the science. My job was to represent lay readers: identifying passages not easy to follow, parts where a slowing of pace, or the introduction of an additional example would help (Richard seemed to have an inexhaustible supply of amazing animal behaviour stories), and so forth. The chapter with the greatest challenge, from the point of view of the general reader, was 'Immortal coils', Chapter 3, which explained the nuts and bolts of DNA, chromosomes, and genes: the original version was too technical for a completely lay readership. On 5 April, Richard telephoned. He was going over my notes on Chapter 3, had started rewriting and while he thought the revision was on the right lines, might he come round to check that I agreed?

What followed is pretty unimportant in the grand scheme of things but it made an impression on me and it has stayed in my memory. It was a hot, sunny day and the window of my office, over-looking the main quad of the OUP building, was wide open. Richard arrived in the middle of the afternoon and suggested that perhaps the best way of proceeding would be for him to read some of his revised passages. It was the contrast between the style of delivery, completely neutral, and the words, which were anything but, that made the episode so memorable. Occasionally I asked him to clarify or expand on a particular point and he would explain, giving the impression that he thought the imparted facts were commonplace. They were small details, certainly, but rather wonderful nevertheless and I urged him to incorporate them. I can remember just one example, tiny, but it gives the flavour. Richard was explaining how plants often propagate by means of suckers. 'You mean, a whole wood of elm trees could be a single individual?' 'Yes.' 'I think it's worth putting that in!'

A month or so later Richard raised the question of having a foreword written for the book. Who might write it? The originators of the main ideas being described in *The Selfish Gene* were Bill Hamilton, at Imperial College London, John Maynard Smith, at the University of

Sussex, and Bob Trivers, at Harvard University. A foreword from any of these giants would be wonderful but one from the American would carry extra weight with the OUP in New York, who would be selling the book in potentially the biggest single market. We discovered that Trivers was on sabbatical leave from Harvard, carrying out fieldwork in the depths of Jamaica, but contact was made and a copy of the typescript despatched to him there. The timetable was tight: with our planned publication date set for October we needed to have the foreword, if Trivers agreed to write one after reading the typescript, by the end of June. Richard received a letter from Trivers in the middle of June: '*The Selfish Gene* has reached me in Jamaica . . . I've thoroughly enjoyed reading your book. From start to finish it is well written and well argued. I hope it gets a very wide readership. It's especially gratifying that the first popular account of the new social theory should have been written by a scientist with a full grasp of the relevant theory and a sure feel for the most telling illustrations. I only hope you can get the book out in time for my fall course. It will be a pleasure to write a short foreword and I should have it to you by the end of June.'

Handwritten, the foreword arrived in the nick of time in early July. It was typed in the office and sent off to the typesetter on the same day. A covering letter from Trivers requested that 400 copies of the book should be shipped to Harvard as soon as they were printed in order to be ready for his fall introductory course.

Meanwhile, in early June, I had written personal letters to the managers of the OUP's branch offices in Canada, South Africa, Australia, and New Zealand. Desmond Morris had stressed to Richard at an early stage the importance of having a publisher who would push the book initially so hard that it would start being talked about. But this was nothing to do with that, it wasn't a carefully orchestrated plan to promote the book throughout the OUP. I had been swept off my feet and my spontaneous reaction was to want to share my enthusiasm, whether this was in chatting to my author Norman Gowar at his local pub in Long Crendon or, now, writing to the OUP branch managers. They would find out about *The Selfish Gene*, and many other titles coming in the same season, through the standard communication channels, but I urged them not to conclude that this was 'routine, or even slightly better than routine popular science'. Then followed my attempt to persuade them: 'This is *not* some worthy attempt to try and popularize an area of science. Forget about science, popular or otherwise, and just think of this as a book that is so readable, so gripping, and so fascinating that, cliché or not, you won't be able to put it down. And I don't just mean you. I defy you to find anyone in your building—accountants, secretaries, salesmen, packers, editors, the lot—who will not find the book fascinating. How many books can you say that about?'

In due course, sets of proofs went out so that people could judge for themselves and an out-of-the-blue response from David Cunningham, manager of the Australia branch office in Melbourne, was for me both heart-warming and astonishing, in equal measure. It was a request for a quote to supply them with 3,000 copies of the book and it ended with a plea for us to tell him if we thought they were 'mad' for contemplating such a staggering quantity. I had had some contact with David at the OUP in Oxford before his move to take charge of the Melbourne office and had him down as the quintessential Mr Caution. It was nice to have been proved so utterly wrong on that particular score.

The book was presented to the sales reps at the summer sales conference at the beginning of July and an issue which had been raised previously was aired, briefly, again: should *The Selfish Gene* be illustrated? Would pictures make this important book even more attractive from the selling point of view? The question revolved for me around what people get out of reading a popular science book. In many cases there is a genuine desire to find out more about a subject, a subject that seems on the face of it to be intrinsically fascinating, say. I want to know more about it, says the reader, and experience some of its intellectual buzz. Popular science books of this kind, and there are many wonderful examples, have an overt didactic purpose, recognized by author and reader alike, and in these cases pictures, including diagrams, are positively helpful—along with top quality writing, needless to say. *The Selfish Gene* was outside this category, or so it had seemed to me when

I first read it. It had felt like reading a novel, appealing wholly to the imagination, with no accompanying conscious wish to learn about a new field—though that was happening, delivered in spades. Having illustrations would dilute the impact, I thought, and lessen the intensity of the experience of reading the book.

Peter Jones rang from the BBC in the middle of September to let me know that the transmission date for *Horizon*'s 'The Selfish Gene' had been set for Monday 15 November, with a repeat on the following Saturday. This meant that the programme would be going out just over two weeks after the book had been published: the publication date had earlier been fixed for 28 October. What would be the effect of the television programme on sales of the book? For us this boiled down to a simple, specific question: should we order a reprint of the book before publication? The dilemma was a familiar one: a reprint would need to be put in hand before the programme went out. If we went ahead with one, but the effect of the film on book buying was minimal, then we would be left with unsold stock in the warehouse; the opposite would mean the book would go out of stock at a critical time.

John Lord, Martin Cowell (the OUP's home sales manager), and I arrived at the BBC's Television Centre on 1 October for an advance viewing of the film. Afterwards we got into Martin's car, having resolved to come to a decision on a reprint before we parted to go our separate ways, and drove around Shepherd's Bush whilst we agonized. Back to our starting point and still no decision. John issued an order to Martin: 'Drive round Shepherd's Bush again.' More agonizing, and yet a third circuit of Shepherd's Bush, but finally a decision: a reprint of 4,000 copies would be put in hand immediately. It turned out to be the right decision.

An advertisement linking the book and the *Horizon* film, to be carried in some of the Sunday newspapers on 14 November, the day before the transmission date, was prepared. It read as follows:

> Konrad Lorenz, Robert Ardrey and others have popularized the theory that animals behave for the good of the species. They do not: they behave for the good of their genes, whose world, as Richard Dawkins shows, is

a world of savage competition, ruthless exploitation, and deceit. This important, controversial issue is the theme of BBC2's *Horizon* programme tomorrow night: see it, and read the book.

But it was not to be. Peter Jones rang me on 5 October to give me the bad news. He had just had a session with his boss, not the editor of *Horizon* but someone more senior, and the boss had only just heard about our plan to link the book with the television programme in an advertisement. The boss was strongly opposed to any such linking, saying that it reeked of 'collusion', and was even prepared to take the ultimate step and cancel transmission of the film if we went ahead with the planned advertisement.

The matter was resolved amicably, and a modified advertisement agreed, in a discussion between John Lord and the new editor of *Horizon*, Simon Campbell-Jones. John summarized what lay behind the drama in a note he sent me, reporting on his conversation:

> From what Campbell-Jones had to say, it would seem that his 'bosses'—he mentioned no names—are as much concerned about the timing of the ads as about what goes into them. Apparently, it's alright for a commercial concern to mention a BBC programme in advertising once the programme has been broadcast (and its audience rating forever incapable of being affected by the additional publicity!), but quite another if this is done beforehand. It is the evidence that that affords of the BBC favouring some outside bodies with details of its programming before the information has been made generally known (e.g. by announcement in *Radio Times*) that worries the top brass.

Peter Jones told us a few weeks after the programme's transmission that the audience had been about 1.5 million. Rather low for *Horizon*, he said, going on to explain that although the programme had been up against the highly advertised *Royal Variety Show*, the opinion at *Horizon* was that the film's title had lost them a potentially larger audience. The with-hindsight view was that the title had put people off because having 'gene' in any title implies hard science and a laboratory-based programme. From the book's point of view, the title was of course perfect.

Oxford University Press in New York were printing their own copies of the book, their publication scheduled for December, but in the meantime 400 copies of the UK edition had been shipped to Harvard to be in time for Trivers's introductory course. These had quickly sold out and Harvard had ordered a further 200 copies. I was visiting New York at the time, in late October, and had arranged a flying visit to Boston, principally to meet Bob Trivers for the first time. Bob's course was already in full swing, with the latest of his lectures timetabled to be given at the end of the morning of my visit, and I asked him if I might sit in on it. The class was large, several hundred students, and what was especially satisfying for me was hearing Trivers announcing at the end the reading assignment to be carried out before next week's lecture: a couple of chapters from *The Selfish Gene*.

The New York biology editor of the OUP at the time was Bob Tilley and Bob travelled to Boston with me. He had spent most of his editorial career at Columbia University Press and whilst there had enjoyed a long association with Richard Lewontin, now a senior academic at Harvard. Bob had arranged to call on Lewontin during our visit and he invited me to join him. Almost a quarter of a century later I commissioned Ullica Segerstråle's definitive account of the sociobiology controversy, *Defenders of the Truth: The Battle for Science in the Sociobiology Debate and Beyond*, published by the OUP in 2000, but at the time of my visit to Harvard in 1976 I was not fully up to speed with the finer points of this battle and its *dramatis personae*. An amusing incident in Bob Trivers's office had provided me with a small insight, however. Bob's phone had rung just before we were about to leave for his lecture. Bob picked up and his short responses during the exchange communicated nothing but courtesy and a desire to be helpful. Putting the phone down, Bob looked over and identified the caller for me: 'Lewontin,' he said. Then, with a grin, he spat out with a pantomime delivery: 'The snake!'

The sociobiology controversy had begun in the summer of 1975 when Edward O. Wilson, a distinguished Harvard entomologist, published his *Sociobiology: The New Synthesis*, in which he defined sociobiology as a new discipline devoted to 'the systematic study of the biological basis of all social behaviour'. The book was soon subjected to intense criticism because Wilson included our own species *Homo sapiens*, devoting his final chapter to humans, suggesting that human sex role divisions, aggressiveness, moral concerns, religious beliefs, and much more, have a genetic basis. In one dramatic episode, some three years after the book's publication, Wilson, about to speak at a symposium sponsored by the American Association for the Advancement of Science, had a jug of water poured over his head by a group of hecklers. Ullica Segerstråle interviewed all of the central participants in the controversy for her account. 'The characters in my story,' she wrote, 'are all defenders of the truth—it is just that they have different conceptions of where the truth lies.'

Lewontin was in good humour when Bob Tilley and I arrived at his office. He told Bob with evident delight that the journal *Nature* had just sent him a copy of *The Selfish Gene* to review. Bob explained that I was the book's editor and, simply to be polite, I said something along the lines of my hoping that he'd enjoy reading it. 'You don't expect me to be nice about the book, do you?' was the response. He then turned back to Bob and changed the subject. The review in *Nature* did not appear until the following March and it was critical in the extreme. Bill Hamilton, one of the giants whose work is described in *The Selfish Gene*, wrote to *Nature* about the review and his letter was published a couple of months later. It began:

> Lewontin's review of Dawkins' book *The Selfish Gene* is a disgrace. It fails to meet any of the standards of informative value, objectivity and fairness to the views of others that are part of the code of science. That it is a very biased review is reasonably evident to anyone reading it; unfortunately a reader unacquainted with the controversy which is its background may well be left with the impression that, even setting aside obvious unpleasantness in the review, the book itself probably is unsound and not worth reading. This is a great pity since in fact the book is not only the best existing outsider's introduction to a new paradigm and a new field of knowledge but, in its overview of the situation and in many original details, is itself a significant contribution to this field.

In March 2006 I attended an event, organized by the OUP and the London School of Economics, entitled The Selfish Gene: thirty years on. During the dinner at the end of proceedings Richard Dawkins gave a speech in which he included a brief anecdote involving me. I had heard it at my retirement party towards the end of 2003: my boss, Louise Rice, had emailed various people requesting stories for her to read out at my party. Richard's anecdote was later stuck inside my retirement card and so I can reproduce it:

> Soon after *The Selfish Gene* came out, I gave a Plenary Lecture at a big inter-national conference in Germany. The conference bookshop had ordered some copies of *The Selfish Gene*, but they sold out within minutes of my lecture. The bookshop manageress swiftly telephoned OUP in Oxford, to beg them to rush an additional order by airfreight to Germany. In those days OUP was a very different organization, and I am sorry to say this bookseller was given a polite but cold brush-off: she must send in a proper order in writing, and, depending on supplies in the warehouse, the books might be shipped some weeks later. In desperation, the bookseller approached me at the conference, and asked if I knew anyone at OUP who was more dynamic and less stuffy . . .
>
> . . . I telephoned Michael in Oxford, and told him the whole story. I can still hear the thump with which Michael's fist hit the desk, and I remember his exact words. 'You've come to the right man! Leave this to me!' Sure enough, well before the end of the conference, a large box of books arrived from Oxford.

I can remember the telephone call from Germany, though not my hammering of the desk, but have no recollection of how I managed to bring about the happy outcome. My guess is that I rang Martin Cowell in the sales department and that he did the rest, but it's a nice story and I was pleased that it had obviously lodged itself in Richard's memory circuits.

Also in my retirement card is a message Louise Rice extracted from Richard Charkin and this can bring *The Selfish Gene* tale to a close. Richard and I were colleagues at the OUP in those days.

> The first print run of Dawkins's *The Selfish Gene* was set at 5,000. I thought we'd be lucky to sell 2,000. Michael was to pay me £1 for every thousand below 5,000 and I was to pay him a pint of beer for every thousand over 5,000. I think I owe him more than he could drink and am holding back payment in the interests of his health and well-being.

CHAPTER FOUR

THE ORIGINS
AND EVOLUTION OF THE
COLLEGE SCIENCE TEXTBOOK
AND THE
BIRTH OF A SUPERSTAR

THIS CHAPTER is about textbooks, and largely about one in particular, a textbook of physical chemistry. There, I've said it, and already I sense that some readers are wondering whether to give this one a miss. Could there really be anything about a physical chemistry textbook capable of capturing the imagination of the general reader? The answer is actually yes: there is in fact an interesting tale to be told about how this one came about. Let me now convince you.

In 1982, I reviewed the book *Wiley: One Hundred and Seventy-Five Years of Publishing* for the journal *Nature*. Being an official company history it was written in an earnest, respectful style but I found one of its strands absorbing: the evolution of the college textbook. 'College', a generic term first coined in North America to mean post-secondary education, was adopted by the publishing industry to describe the part of the business devoted specifically to textbooks. The story of the college science textbook as big business essentially began in the 1940s when the American government enacted its GI Bill of Rights. This entitled veterans of the Second World War to educational assistance, thus boosting the demand for college textbooks that had begun in the early 1940s, which in its turn arose from the armed services' large-scale provision of technical training. The bonanza for American textbook publishers came to an abrupt end in 1948, and not simply because the last of the ex-GIs had completed their college education. The boom had fostered complacency, since little effort was required on the part of publishers to satisfy returning GIs who were prepared to put up with dull, old-fashioned textbooks; their successors straight from school were more particular. Textbook publishers found that they would have to mend their ways, and gradually they did.

Textbooks in the early days were the result of the buckshot approach: the publication of books with a broad appeal that *might* be adapted to specific course needs. In the 1950s the search for the customer began after a textbook had been produced. In those days the role of basic introductory textbooks in providing a reliable and substantial backlist income for the publisher was yet to be fully appreciated. Things began to change during the 1960s and the 'development' of science textbooks became more professional: in the United States the era of College publishing had begun. A pair of figures relating to a shift in Wiley's share of the college market during the 1960s rubs the point in. In 1961 their list contained 18 college titles with annual sales of 10,000 copies or more; by 1971 the figure had risen to 46.

The Wiley company history provides a fascinating glimpse of the genesis of a famous science textbook published in 1960, Halliday and Resnick's *Physics for Students of Science and Engineering*. What a prosaic title! Why did it become famous? All mainstream science textbooks have prosaic titles—they are simply the titles of the course for which they have been developed—but the books themselves are never referred to by their titles, but by the names of their authors. *Halliday & Resnick* became famous because it sold in prodigious quantities, and thus made real money for the authors and the publisher.

How do the really successful textbooks, and *Halliday & Resnick* is one, actually begin life? It is a chancy business, for all of the talk there might be about careful planning, with the right author and the right editor coming together just when the market is ready for what they will eventually develop. Towards the end of 1954, one of Wiley's college editors, Robert B. Polhemus, happened to be visiting Resnick at the University of Pittsburg. A number of publishers had been urging Resnick and his departmental colleague Halliday to turn their lecture notes into a textbook. Polhemus noted this in his report to Wiley's New York office, continuing:

> The idea has blown hot and cold. Right now it's cold, and Halliday told me afterwards that he has definitely decided against doing such a book.

But Halliday and Resnick did press on and the rest, as they say, is history. Published in 1960, the parent text and its various abridgements and supplements had sold by 1982, when the Wiley company history was published, almost two million copies.

No textbook is perfect, and an author thinking about writing a new one should begin by dissecting the competition, identifying each rival text's shortcomings, while the publisher's editor will research what the market—lecturers who assign for purchase a textbook to accompany their course—would like to see that existing texts don't provide. Completed draft versions, either of individual chapters or the whole thing, go out for 'review', meaning that they are sent out to selected lecturers who teach the course, and their comments and suggestions are evaluated by author and editor during the next revision stage of the writing. Big introductory courses mean large potential sales for the right textbook, in turn meaning fierce competition, with rival textbooks jostling to secure adoptions. For these high-stakes markets enormous trouble is taken to maximize the probability of getting things right, one example of this the assignment of a 'developmental editor' who will work through a typescript line by line to ensure that every sentence could not be clearer. Strange but true.

So much for the writing and the development of a textbook. What about selling it? Here, selling means convincing college professors that the new text, though it might not be perfect, is better than the existing rivals. If it is, it will be adopted for that course. It was not until the end of the 1980s that I saw at first hand how the very best of the textbook selling campaigns waged on American campuses are planned and executed. This was when I attended the Benjamin Cummings and Addison-Wesley college sales conference in Tucson, Arizona, in January

1989. I was then with Longman, the British sister publishing house of Benjamin Cummings and Addison-Wesley, all three owned at this time by Pearson (see Chapter 7). For me, as an outsider, a visitor from the UK, it was an amazing experience and I was completely bowled over by the professionalism that had plainly gone into the planning of every aspect, and the dazzling showmanship on display in the major presentations. The purpose of everything taking place was deadly serious—to provide the sales force with all the ammunition needed on their campus visits to persuade college professors that each new textbook being presented was worthy of a course adoption—but the atmosphere was good humoured and friendly throughout the entire week. Being there was certainly riveting for much of the time, but it was fun too.

The Benjamin Cummings contingent had about them the sort of easy-going presence you might expect from a small publishing company based in San Francisco. Their main textbook focus was biology. Addison-Wesley, Boston-based and bigger than its Californian sister, had more of a corporate formality about it. Its publishing was centred mainly on physics and engineering. Both companies had a computer science list, the one area of overlap, and their rivalry was amusingly revealed during a small Benjamin Cummings presentation of a forthcoming computer science text I was sitting in on. A list of the existing rival texts being targeted included an Addison-Wesley title and, when its turn came, its weaknesses were clinically exposed. The enemies' vulnerabilities identified, the presenter issued a blunt declaration of war: 'Let's kill 'em.' A couple of besuited corporate types from Boston looked visibly startled, bemused that the battleground extended into their backyard.

The high point of the sales conference came at its beginning. Benjamin Cummings had been allocated two of the seven days and they lavished the entire first morning, a quarter of their time, on just one book: *Human Anatomy and Physiology* by Elaine Marieb. Written and developed for big, high-enrolment courses, and being launched against stiff competition, it had a great deal riding on it. At the opening plenary session a series of presentations left us in no doubt about the book's importance, or about the care that had gone into its development. Then dispersal for so-called breakout meetings at which sales reps learned about every conceivable selling feature of the book and about how to present these during their campus visits. And finally back to the main auditorium for the concluding plenary session. A link between the book's subject, human anatomy and physiology, and the achievements of athletes had earlier been made concrete in our minds by associating it with a famous event, the women's 4 by 100 metres relay race at the previous year's Olympics at Seoul, in 1988, and in particular with one of the athletes who had run in the race: Evelyn Ashford. On the cover of the book was a stylized picture of Ashford, running in a swimsuit, shot by the American portrait photographer, Annie Leibovitz. This inspiring connection was to be reiterated and then hammered home in the final session: we were to watch a film of this justly famous race. It is famous on account of what happened in the final stage, the anchor leg, run for the American team by Evelyn Ashford. There was a messy baton exchange and Ashford began three metres behind the leader, the East German competitor, but Ashford gradually overtook, her astonishing run winning gold for the Americans. Watching the film was genuinely thrilling and there was much shouting and screaming. A fitting climax to the launch of *Marieb*, we all thought. But the Benjamin Cummings showmanship machine had not yet finished with us. As soon as the film had ended an announcer at a podium on the front stage spoke into her microphone. 'Please welcome' —her voice then ratcheted up for a shouted, theatrical delivery— '*Evelyn Ashford!*' The earlier shouting and screaming were nothing in comparison as the track-suited heroine appeared from nowhere, cantered down the central aisle, mounted the stage and delivered what came over as a spontaneous rousing speech, telling us how the effort that had gone into *Marieb* reminded her of the training behind her running of the anchor leg at the Seoul Olympics; and then finally she urged the sales force to 'Get out there and sell this great book.'

Hats off to Benjamin Cummings. Were their mighty efforts rewarded? Did the book live up to its promise? Twenty years later, in 2009, the seventh edition of *Marieb* was still selling.

In 1969, when I started my job as a science field editor, I knew little if anything of how textbooks were developed and sold but I began to wonder whether the OUP might expand its chemistry textbook publishing. I called on numerous chemistry lecturers during my campus visits to find out what books were being used, what was wrong with them, and what people would like to see. The term 'horizontal approach' was being bandied around and sold to me as the Next Big Thing in chemistry teaching. The traditional divisions of chemistry, physical, inorganic, and organic, were being broken down, enthusiasts assured me, and the teaching of the future would involve exploring how individual chemical concepts interacted with the historical physical, inorganic, and organic divisions: the horizontal approach. Might this provide an entrée into the chemistry textbook world? Eventually I called on R. S. Nyholm, Professor of Chemistry at University College London. An Australian, Nyholm was a noted inorganic chemist with a special interest in the teaching of chemistry in schools. I told him what I had learned about the horizontal approach and about how I thought a new series of textbooks on individual topics in chemistry might be welcomed. Did it make any sense? Or as we might put it these days, was the idea likely to fly? It was lunchtime and his response was encapsulated in his next sentence. 'Young man,' he said, 'I'd like you to join me in a glass of sherry.'

The episode remains a pleasant memory but the reason for recounting it is that it was the first time someone of Nyholm's stature and importance had supported so positively the idea of the proposed series. Nyholm subsequently sent me a letter confirming this and I showed it to Rex Richards, Professor of Physical Chemistry at Oxford. Rex was supportive and said he'd like me to come round to the Physical Chemistry Laboratory, the PCL, to meet some of the physical chemists and tell them about these ideas. The visit was arranged for the afternoon of 15 June 1970 and I went along expecting that I would be introduced to a few of Rex's colleagues and that I would chat to them informally about the proposed series. I was in for a shock. The Oxford physical chemists, I think there must have been around twenty of them at that time, took tea each day at four o'clock in the PCL's tea room.

On my arrival it was immediately announced to all that 'Michael Rodgers has come round to tell us about the OUP's plans for chemistry textbook publishing'. Totally unprepared for this, I faced my audience and immediately picked up signals that, for reasons about which I was totally ignorant, there seemed to be some sort of animus towards the OUP. Certainly the atmosphere was a trifle tense as I ploughed on with my unprepared address, and I concluded that the PCL was surely a lost cause. Yet I was wrong. When I'd finished and there was an invitation to ask questions, two of the physical chemists—one of them being Peter Atkins—made constructive comments. Peter was certainly well-disposed towards the Press—his textbook, *Molecular Quantum Mechanics*, had been published by the OUP in the previous year—and I later arranged to meet him to talk about whether he might agree to become involved in a possible chemistry textbook series, as one of its editors. Eventually Peter said yes: it was to be the beginning of a publishing association with him that was to last for more than thirty years.

At around the same time as my baptism by the Oxford physical chemists I visited Bristol University where one of my meetings was with Mark Whiting, professor of organic chemistry. I launched into my standard spiel on the horizontal approach and how closer coopera-tion between chemists from the different branches of the subject seemed to be on the cards for the future. Whiting was dismissive, telling me that at Bristol they didn't require lectures on departmental bridge-building. Why, only this morning, he went on, I received this note from my colleague, David Everett (Bristol's Professor of Physical Chemistry), handing it to me so that I could read it for myself. In essence the message read something like: Dear Mark, Yes, we really must get together more often to discuss matters of mutual interest. Yours ever . . . Whiting looked over at me triumphantly, the delicious irony apparently having escaped him. Their offices were located in the same building and yet the card bore a postage stamp: it had been sent by Royal Mail.

The embryo series was given a name, the Oxford Chemistry Series, to comprise single-topic paperbacks of a hundred or so pages each and selling for £1. Two more series editors, both at Liverpool University, were signed up, Ken Holliday, Professor of Inorganic Chemistry and author of some school-level textbooks, and the organic chemist Stan Holker. They joined Peter Atkins to plan the series and shortly after the proposal was approved by the OUP's Delegates. It was an exciting period, with regular planning meetings with the three series editors every six weeks or so, and there was a satisfying feeling as things quickly took shape. There was more of a buzz when Adam Hart-Davis joined the Press as college science editor in the late summer of 1971. Adam brought his energy and infectious enthusiasm to the running of the series and it was fun working with him.

At one of the series planning meetings, Peter Atkins suggested it would be useful to seek the views of Fred Dainton on a volume we wanted on chemical kinetics, and I volunteered to go and see him. I had first met Dainton at a reception shortly after his appointment as Professor of Physical Chemistry at Oxford in 1970. It had been an interesting move: Dainton had spent the previous 15 years as Vice-Chancellor of Nottingham University and, during that time, he had chaired a Government enquiry into the decline in the numbers of university entrants in science and technology, the Dainton Report on 'The Swing away from Science', published in 1968. My meeting with him had been arranged for Saturday morning, 13 May 1972, at the Physical Chemistry Laboratory. The PCL was all but deserted when I arrived but I could hear Fred's familiar voice coming from his first-floor office. He was with a student, patiently going over a point of chemistry, and I stood in the open doorway until he'd finished. Eventually the student signalled that he now understood what had been puzzling him, and left. Dainton greeted me with a broad smile, unable to conceal his delight. He had put himself down to teach an introductory undergraduate course and had told his students that his door would always be open if any of them needed further explanation. A student had taken him at his word that Saturday morning and this gave Dainton enormous pleasure.

I remembered this incident some 25 years later when Dainton, by then Lord Dainton, had died. The obituary in the *Independent* was by Tam Dalyell and he made the point that the professorship of physical chemistry at Oxford was a relatively restrictive role for someone of Dainton's range and influence by then. That was certainly a widely-held viewpoint but I had glimpsed another side to him. The *Independent* published my post-obituary letter in which I described the Saturday morning episode: his continued commitment to, and enjoyment of, his teaching duties were a real tribute to his character and achievements.

The Oxford Chemistry Series was launched with the publication of its first four titles in July 1972. Its final volume, the fourth edition of Brian Smith's *Basic Chemical Thermodynamics*, title number 35, was published in 1990, eventually going out of print in 2004. (And since picked up by Imperial College Press for its sixth edition.) And two series titles, including Peter Atkins's *Quanta*, are available in a print on demand format and so technically they will remain in print indefinitely.

Was the series a success? Had it been worth doing? One indisputable result was that it swiftly raised the OUP's profile in chemistry textbook publishing. But in the science department around the time of the OCS launch we began to debate the Small Grocer Problem: making a living from selling lots of different, cheap items. Beginners on the science side of publishing tend to be fond of series: they are relatively easy to start, and early on there is the exhilaration that comes from feverish activity and a feeling that steady progress is being made as successive titles are commissioned. When I read the Wiley company history for my review in *Nature* in 1982 I smiled when I came to a comment from the 1950s, the era just before the development of the big course textbook began in earnest: 'Editors relied on advisers, series editors, consultants, boards, and so on. Just look at our catalogs of those years. Nothing but one series after another.' I smiled because at the OUP we had started out doing much the same thing, though twenty years later.

It had begun to dawn on me around the time of the launch of the Oxford Chemistry Series that real success, meaning substantial sales income efficiently generated, could come only from One Book, a textbook carrying a much higher price tag than the £1 of an OCS volume, and selling into a much bigger market.

This vague aspiration was eventually to become reality but before moving on to that I want to insert a postscript on the much-vaunted (at the time) horizontal approach. This, after all, was the essential catalyst during the initial stages when the OCS was conceived. I had all but forgotten about it at the time of a routine campus visit to Sussex University at the beginning of 1976 but the memory came back on discovering that the Sussex chemists called their department the School of Molecular Sciences, designing their courses around a series of interdisciplinary units: the horizontal approach, no less. The concept had in the event played very little part in our discussions on the individual titles we decided to commission, but that didn't matter: it had done its job in getting things started.

A group of academic booksellers had been invited to attend part of an OUP sales conference in 1977, and I was asked to give a short talk on the story behind *Physical Chemistry*, a new textbook by Peter Atkins due to be published the following January.

In my opening sentence I said that I had asked Peter if he would write the book on Monday evening, 10 December 1973, at five past seven. Wanting to keep the mood light, I guessed that this precision would be taken as a tongue-in-cheek invention, but given five or ten minutes either way, it was in fact spot on. A sentence or so later I went on to say that the first hint of the need for the new book came to me on Tuesday, 17 February 1970, at twenty past four in the afternoon, and given five or ten minutes either way this again was accurate. That said, I didn't know at the time that this particular hint was to mark the beginning of a story: I pieced it together much later. I was on a routine campus visit talking to one of the chemists at Hull University and he happened to mention in passing that he and his colleagues were fed up with the existing physical chemistry texts. Other people said the same thing over the years and gradually the idea that there was a genuine need for a new text grew. What happened next, towards the end of 1973, came about by pure chance. The OUP's New York office had signed up a proposal for a physical chemistry text years earlier and when the draft typescript was delivered it was sent out for review in the normal way. The American office knew about Atkins through the Oxford Chemistry Series, which they were trying with difficulty to sell in their market, and they chose him as one of the reviewers for their new text. I knew nothing about this but—the pure chance bit—a copy of the Atkins review accidentally landed on my desk. The review was long, detailed, and extremely critical, and it told me more about Atkins than about the typescript being reviewed. For me it seemed to be blindingly obvious, with no great insight on my part needed, that the new textbook should be written by Atkins. He had earlier invited me to dinner at his college, Lincoln, and that was when I popped the question, over our preprandial drink on that December evening in Oxford in 1973.

Peter took his time before deciding. What worried him, he said, was the prospect of gaining a reputation for being able to write only textbooks, and thinking of his professional career, he wondered whether he ought first to write a scholarly monograph. But he didn't say no. The idea had been planted, he said, and it should be left at that for the time being. He agreed early in the new year to go ahead with his attempt and the first thing I did was to buy in copies of the competition. The collection of substantial tomes was duly delivered to Lincoln College and I asked Peter to read all of them, cover to cover. The main rivals in existence were (remembering the convention that mainstream science textbooks are referred to by their authors' names, not their titles) *Moore*, *Castellan*, *Barrow*, and *Daniels & Alberty*. I've listed them in descending order of level, as we then saw this: *Moore* was judged to be the most demanding, and as the leading text in the UK we saw it as our main target; *Daniels & Alberty* was regarded as the easier of the rivals, from the point of view of students. Peter wanted to pitch his book, in terms of degree of difficulty, at an intermediate level, between *Castellan* and *Barrow*. And finally on this, all of these texts, and some lesser rivals, had one striking thing in common: they were all American. A widely held view at the time was that it was not possible for a British author to write a science textbook that would succeed in the American college market. This was because it was felt, with good reason, that direct experience of teaching physical chemistry at an American university, where the course closely follows the assigned textbook, was essential for writing a successful textbook for that course.

Whilst Peter was working his way through the competition I carried out a preliminary market survey. This involved writing to a number of academics and going to see some others. The exercise confirmed that the need for a new textbook was real, and it also identified some key points on how a newcomer should differ from the existing rivals.

The evaluation of the competition and the market research both completed, we decided on our plan: three specimen chapters and a contents list for the whole book would be prepared by the end of September, 1974, and sent out to selected academics for criticism and comment. We chose the reviewers together, four from the UK and one from Australia. The OUP's New York office independently lined up two Americans.

The reports came back on schedule by the beginning of December and the result was unanimous: we'd got it completely and absolutely wrong. In a sentence, the style of the writing was too informal and chatty, and the mathematical level was too low. Over lunch, Peter and I met to decide what to do. We went over each report, occasionally wincing as we reread passages of particularly harsh criticism, but gradually we began to convince ourselves there were hints that we might be on to something. I ordered a second bottle of the red and the discussion went on. It was eventually decided that Peter would start again and completely rewrite the chapters. We selected the two reviewers whose criticism had been the most constructive to read the revised chapters, and the OUP in New York organized two new Americans for the job.

The revised chapters were sent out in March 1975. And the verdict this time? In a word, inconclusive. One reviewer thought that while a lot was wrong there was promise: we should press on. Another thought we should give up. Two sat on the fence. The OUP in New York concluded that on this evidence they didn't believe it would work, and given what was in front of them, that judgement cannot be criticized.

It was decision time again. Our spur was the eventual realization that no reviewer was ever going to predict success or otherwise on the basis of ten per cent or so of the whole thing. So Peter decided to write 22 of the projected thirty chapters (two of the three sections of the book) before any more reviewing. It was agreed at the Press that I should try to interest another American publisher in the book, and that the OUP in New York should be informed about this at the outset. I made contact with W. H. Freeman in San Francisco, a prestigious textbook publisher without a physical chemistry textbook in its list. Their reaction was sceptical but they agreed to have the 22 chapters evaluated when these were ready, scheduled for completion by Peter for January 1976. A decision was promised by early June.

Peter sent me the draft of each completed chapter and there was scope for me to wield my editor's pen. On the one specific instance I can remember, a chapter on spectroscopy, there was much to criticize and I returned it covered in red ink. A note of acknowledgement from a wounded Peter told me that the chapter had brought out the schoolmaster in me, but Peter always took criticism on the chin and if it was valid he was glad to learn from it. I hasten to say that my points were not technical ones on the chemistry but rather were concerned with presentation, how quickly or slowly a new idea is developed, and so on. I am mentioning this here because it now seems so strange that it happened, given that Peter went on to become one of the most famous and gifted textbook authors in the world. It is nice to be reminded of a time when we did things in this homely fashion.

The 22 chapters were duly delivered on time in January, to be copied and sent to San Francisco, and I can remember that this was the moment when I thought we had a winner: such was the dramatic transformation from the drafts I'd seen before to the beautifully clear revised version I quickly scanned when Peter handed it over to me.

Towards the end of April I received a note. Handwritten on Lincoln College letterhead and dated 22 April 1976 it had at the bottom of the sheet a dot surrounded by a border. Above this the note read: 'Dear Michael, In the illustration below you will see an almost exact facsimile of the full-stop at the end of the final chapter of the little thing I am doing for you. The original was created on Tuesday evening. I am sure you will agree that this full-stop is a milestone. Yours sincerely, Peter.' I sent a reply: 'Dear Peter, Thank you for your letter. In the illustration below you will see the marks made by two appendages which will be permanently crossed until I hear from some people in San Francisco I've been in touch with about the little thing you are doing for me. Yes, of course I agree that the last full-stop is a milestone. But there must be no extravagant celebrations at this stage, lest we tempt Fate!'

The first American review arrived from W. H. Freeman in May and it was very positive. Two more followed, both equally enthusiastic. We then knew we had a textbook that could succeed in the American college market, and this in turn meant it could do the same anywhere in the world. It was then, and only then, that the book was signed up.

The best time to publish a mainstream course textbook is in January, allowing maximum time for promotion—aimed at securing adoptions for the academic year starting in the following October—while at the same time allowing the book to incorporate that same year as its year of publication for an entire 12-month period. Peter undertook to complete everything and deliver the typescript on New Year's Eve 1976, which would give us a comfortable schedule with a publication date in January 1978. For the rest of 1976 he revised his text, incorporating suggestions from the American reviewers, and composed over a thousand end-of-chapter Problems. Typically for Peter, delivery was slightly early, on 23 December.

A diversion when the writing of the book was in its final stages was provided by Peter's chance discovery of two indexes, SMOG and FOG, which produce a measure of the accessibility level of writing in terms of average sentence lengths and relative concentrations of short and long words. In SMOG, you count the number of words of more than two syllables in ten consecutive sentences, take the square root, then add three. The result is an index of level (actually equivalent to the grade of the reader in the American school system able to understand every word). Comparisons rather than absolute values are interesting and Peter compared the SMOG index of his book, which scored 13, with *Daniels & Alberty*, the text we judged to be the most accessible of the competition, which scored 15.

In FOG, you work out the number of words in a sentence of average length and add to this the percentage of words in that sentence of more than two syllables. At the time of writing this chapter, the leader in today's broadsheet *Daily Telegraph* scores 40; the one in the tabloid *Sun* scores 16. As with SMOG, comparing similar things is more illuminating than looking at absolute values, in this case meaning a comparison of physical chemistry texts, because long words are unavoidable in science. Peter compared his first chapter with the first ones in *Moore*, which we saw as the most demanding rival in terms of level, and *Daniels & Alberty*, the most accessible. *Moore* scored 47, the other two, 38. Finally, squeezing out of the exercise just about every comparison possible, the prefaces of the three texts were analysed. Prefaces, it was argued, probably have fewer specialized scientific words than in the body of the books themselves, and so might provide a fairer assessment of an author's natural style. *Moore* scored 48; *Daniels & Alberty*, 43; and *Atkins*, 34.

The exercise was, as I say, no more than a diversion. Peter obviously didn't set out to write with SMOG and FOG in mind, the book was virtually finished before the indexes were discovered, but at the time it was fun doing it, though that of course was only because we found the comparisons reassuring.

We were in for a shock when the galley proofs arrived. Galleys are produced in strips, rather than the made-up pages of the page proofs that would come later, and measurements on the galleys using rulers were hastily carried out to estimate the book's length. At the beginning of the saga, before Peter started writing, all of the advice from the marketplace pointed to an ideal length of 600 pages (all of the existing texts are too long, everyone said). As the writing got under way this target grew in our minds and gradually 750 pages became acceptable as the limit. Certainly it became fixed in my mind as the absolute maximum. Then suddenly the galleys pointed to an extent of just over a thousand pages and I was plunged into gloom. But there was nothing to be done and so we turned our minds to a way of making a virtue of the overshoot. Little effort was in fact needed: we were able to claim that the greater than expected length was not the result of the cramming in of more and more facts, something the market would find difficult to forgive. On the contrary, our promotion literature was to stress, the book was long because it had more illustrations, more worked examples, both of these features designed to help the student. The actual text itself was shorter than the competition.

I had by now moved on from being the science field editor but those first five years in publishing doing that job had stamped on me a belief that the best way of finding things out was to go and see people, talk to them, and listen to their views, face to face. I felt this desire to discover something with my bare hands as soon as the *Atkins* proofs had been produced and I decided that I simply had to take a set with me and visit a variety of university chemistry departments in order to get a direct feeling for how the book was likely to be received. After returning from a multi-campus trip at the end of May 1977 I wrote a note to my colleague Simon Wratten, academic marketing manager, making the point that we 'surely have a book here that is streets ahead of its rivals'. 'Sweeping the market,' I wrote, 'may seem like an exaggeration but considering the wide range of customers I've seen, their almost un-qualified praise, coupled with a general dissatisfaction with the com-petition, it can't be too far from the truth.' Simon and I had discussed the sales forecast the previous summer when the book was being signed up and we had come up with a proposed first print quantity of 5,500 copies. In my note I argued that this quantity was surely 'far too modest in the light of the soundings I've been taking with the proofs.' The nightmare scenario would have us running out of stock in the middle of the book's first year, if the book 'takes off beyond our wildest dreams'. I argued for an increase in the first printing to 7,500 copies and this was eventually agreed.

I talked to many people in chemistry departments at that time, get-ting views based on the proofs at first hand, but one meeting stands out in my memory above all of the others. It was with Professor Max McGlashan, head of chemistry at University College London, during the afternoon of 28 June 1977. McGlashan was a thermodynamicist and the high priest of symbols, units, and nomenclature in chemistry, and he had a reputation as a pedant who conducted arguments with a re-morseless logic. He went through the proofs page by page—it soon became clear to me that this was to be a long session—and he began by picking up points of extraordinary detail. Each of these began with a softly delivered, mournful 'I'm sorry that he . . . ' (an early example being that the decimal points were above instead of on the line) and each was rounded off with his solemnly intoning 'as now internationally agreed'. He lighted on Avogadro's constant, an important quantity in physical chemistry, which is given the symbol L (because, Atkins mentioned in passing, Loschmidt, a 19th-century Austrian chemist, was the first to measure it). It has nothing whatsoever to do with Loschmidt, McGlashan assured me, going on to explain how he knew this: he, McGlashan, had been a member of the international conference taking the decision on the symbol to be adopted, and L was chosen because it was the only available letter of the alphabet left. He drew attention to weightier mistakes and my heart began to sink. But presently he looked up and said that his points of criticism were minor ones; there was plenty that he liked. I then began to realize that the occasional high-pitched humming and the sound of air being sucked in through almost closed lips were manifestations of pleasure. McGlashan became engrossed, said he was enjoying himself, and hoped I didn't mind the time. He followed many of Atkins's arguments, step by step, body language and sounds indicating intense pleasure. But his hawk eyes kept noticing points of detail: I'm sorry that he uses electron volts, or Angstrom units, or Debyes . . . And my own favourite, because in terms of abstruseness it is completely off the scale: the azeotropic diagrams are wrong, as they are in every other textbook, because the touching point should be two cusps, not one. This arcane morsel will be appreciated only by a select band of professional physical chemists but I am recounting the detail here to give you a feeling for the meeting.

It was finally time for McGlashan's verdict. 'A bloody marvellous book,' he said, and guaranteed that we would have the UCL adoption the following year. The book, he concluded, will 'win the market'.

The meeting had lasted for three-and-a-quarter hours. When we descended to the entrance hall we found the department deserted, everyone having gone home. As the doors had been locked, McGlashan had to search for his set of keys. We parted outside and I made my way back to Paddington station feeling exhausted but elated. It had been an unusual baptism and I now felt certain we would secure far more than the adoption at UCL.

We were by now, June 1977, confident that the initial print quantity we were proposing was the right one, but another key variable was still to be settled: the price to be charged for the book. Our main target was *Moore* and during my travels to chemistry departments with the proofs I asked everyone for their view on pricing. The consensus was that, yes, we could charge more for *Atkins* than for *Moore*, it was a better book, but by no more than £1 if we were to avoid hampering *Atkins* with too high a price. *Moore* was an American text but it was published outside the US by the UK-based Longman. Was it possible to discover what Longman would be charging when we published *Atkins* at the beginning of 1978? In the middle of June I popped into Blackwell's bookshop in Oxford and, one of my routines, checked the price of *Moore*, discovering that it had been price stickered at £6.95. This seemed like a recent increase from the £6.25 I was used to seeing and so I contacted a chemistry lecturer friend and asked him to ring the science editor at Longman to find out if the price was likely to stay the same until at least the start of the next academic year. The editor told my friend that the price rise to £6.95 was indeed a recent one, and that it had been approved by the Price Commission and so would be fixed for one year. A reminder here that at that time price rises had to be sanctioned by the UK government's Price Commission, and the new price was then frozen for a year. It was agreed without further ado that *Atkins* would be priced at £7.95 on publication.

Atkins's *Physical Chemistry* was published at the beginning of January 1978 and shortly after became the leading text for physical chemistry courses in the United States and throughout the world, a position it still holds today. A review of the first edition in the American journal *Choice* put its finger on why it was so special:

> Virtually all chemistry curricula include junior- or senior-level courses in physical chemistry. The textbooks used are encyclopedic, everything-the-

author-thinks-a-professor-might-think-should-be-taught compendia decorated with tradition-laden numerical problems, historical anecdote, and museum pieces of chemical knowledge. This book is different from the rest in a special way, in that it was written by an author with very special gifts for exposition and a technical knowledge of modern pedagogy. The coverage and level are the usual, more or less identical to the perennial bestseller *Physical Chemistry* by Walter J. Moore . . . but in Atkins' volume the student is led by a strong hand . . .

For another take on how the world greeted the first edition I want to reproduce the reaction of a Canadian academic. Potential adopters of a course textbook are sent a complimentary copy of the book along with a comment card inviting feedback. Such comments are usually short, unembellished statements of fact, but this one caught my eye because it came straight from the heart, giving it a sweet, over the top quality:

> It is clear that Dr Atkins is an inspired teacher. So many questions that students of physical chemistry ask, the answers to some of which I have managed to puzzle out over twenty years, are answered in this book in a clear and elegant way. I shall recommend this book to all serious students of chemistry, including graduate students. This book makes delightful reading and shall lure me from my other evening reading for some months to come. A lovely source of new problems. A model of writing style for students—God knows they need one.

Peter's labours on his *Physical Chemistry* had in 1978 only just begun, with a new edition published, with clockwork precision, every four years from then on. The nine English-language editions have combined sales to date of more than three-quarters of a million copies, but was the book by now getting tired? Was a new rival waiting to pounce? In 1998, to celebrate the 20th birthday of *Atkins*, the OUP published an article about the book in its in-house magazine and the piece revealed a remarkable fact: every edition had been more successful than the preceding one. The main reason for this was and is the extraordinary care and effort devoted by Peter to the preparation of each revised edition, a point brought home in a review of the sixth edition, published in 1998, in the journal *Education in Chemistry*:

The first edition revolutionized undergraduates' ability to learn/revise physical chemistry and subsequent editions have continued to alter the face of physical chemistry. In some cases new editions bring only small changes to the previous text, often simply correcting mistakes. This is not the case with this book. Each new edition has involved major changes in content, presentation and style, ensuring an enormous contrast between the first and sixth editions.

I began this chapter with a reference to my review of the Wiley company history for *Nature*. In the review I mentioned the merger in 1961 of Wiley with Interscience Publishers, the creation of two Europeans, Maurits Dekker and Eric S. Proskauer. Their original aim had been to provide a European-style publishing house in New York for émigré scientists in America. When the firm was established in 1940, German chemistry texts were used throughout the world: one of the first Interscience titles was an English translation of Karrer's classic *Organic Chemistry*. Thirty years later roles had been reversed, and the translation of Cotton and Wilkinson's *Inorganic Chemistry*, published by Wiley, became the standard text in German universities. To this list can be added Atkins's *Physical Chemistry*: the German translation of each edition has outsold all of its domestic rivals.

The editor of the OUP in-house magazine who commissioned the 1998 anniversary article on the Atkins book called the piece 'The story of a superstar'. I thought it was a fair description of both book and author.

CHAPTER FIVE

A COMPANION TO THE MIND
AND
SCIENCE IN THE
VEGETABLE GARDEN

In 1974, Dan Davin, head of the OUP's academic division, the new name for the Clarendon Press, told me that it had been decided to split science publishing into two separate departments, saying that he'd like me to take charge of one of them. I had by then been the science field editor for five years and while I still loved the job I was increasingly feeling pangs of regret when handing over to my in-house editorial colleagues projects I'd nurtured to the stage when they were ready for formal commissioning. So there was a desire on my part to come inside and take responsibility for my own list; Dan's invitation had come at just the right time. What needed to be settled was how the science list was to be divided. Between physical science and, the part for me, biological science, Dan wondered, or 'another way'? I didn't hesitate to go for the 'another way' option because by then I was quite clear about the sorts of books that most interested me: textbooks, trade (meaning for a general readership) reference books, and popular science. On 1 October 1974 the College, Reference, and Popular Science department was officially born, and it felt wonderful: reporting directly to Dan, the Oxford Publisher, I'd been given the freedom to build my own list. As a department it was by any standard tiny, but that was fine by me. I had no desire to be in charge of an empire, with lots of people reporting to me; more than anything I wanted to work with authors and publish individual books that would stand out from the crowd, sell, and make money.

My interest in trade reference publishing had first been sparked by a memo from Dan headed 'Companion to Science' and circulated at the beginning of 1972 to the science editors for comment. Dan had been reminded of the idea for such an Oxford Companion, something that had been played with from time to time but eventually abandoned because of the formidable problems stemming particularly from the vast range of the subject, and the difficulty that much of the material would quickly date. But he wondered if it might be worth thinking about again. I wrote a response, arguing that such a book shouldn't be seen in terms of an encyclopedia, with lots of narrow headings. Rather, I thought, the headwords should be broad, and a main purpose of the book should be to communicate to non-scientists the excitement of science. In April 1974 I talked to J. G. Crowther about my ideas on a Companion to Science and this, I wrote in my note on the meeting, clarified my thoughts. In the note I wrote: 'When we first began to think about a Companion to Science I had in mind two sorts of approach which I thought could somehow be united. One approach would be concerned with communicating to intelligent non-scientists the excitement, the flavour, and the implications of present-day science; the other (more traditional) approach would aim at providing biographical details of important scientists. As I outlined these ideas to Crowther I began to realize that the first approach was not really on: it would not be possible to sustain it throughout a work of this length, and in any case I think that lively treatments of exciting topics really means individual books (in the context of our popular science programme). I now believe that a Companion to Science should be organized on traditional lines: it should be a straightforward reference book, not a collection of articles designed to excite non-scientists.'

In a response to this note, Dan wrote that what I was now proposing would provide very much what he had always thought a Companion to Science should be. Shortly after this, Dan suggested that I read the original file for Harvey's *Oxford Companion to English Literature*, the OUP's first Companion. The book had been the idea of Kenneth Sisam, then Assistant Secretary (the OUP's second in command) of the Press, in the late 1920s. Reading the file, following the story from genesis to completion—the first edition was published in 1932—was totally riveting. It was especially fascinating to see how the formidable Sisam dealt with advisers who produced endless lists of what might be included, and an anxious OUP office in New York, with their eleventh-hour requests for more entries on American work. Sisam stuck to his guns and trusted Harvey's judgement where selection was concerned. At one stage, in exasperation he cabled New York: 'It is selective and completeness and perfection impossible.' And finally, responding to the New York office's pleas for a delay so that further suggestions might be evaluated, Sisam cabled that the book was going to press forthwith. Reading that file was certainly instructive, and inspirational too.

By the summer of 1974 an editor for the proposed *Oxford Companion to Science* had been found and there followed a period of months when lists, headwords for one letter of the alphabet, for one subclass of a branch of science, were produced together with sample entries. All of this material was circulated and commented on by the science editors, colleagues in sales and in marketing, and in the OUP's New York office; by Dan Davin and even, occasionally, by the Secretary, the Chief Executive, himself; and by the science Delegates. This was a wonderful thing about the OUP, certainly in those days: the sheer range of expertise on tap and a ready willingness on the part of everyone to take trouble in providing thoughtful feedback. I went further afield and enlisted the help of Harold Osborne, editor of the *Oxford Companion to Art*, published in 1970. Osborne happily commented on the lists and the specimen entries and, in February 1975, agreed to talk to the proposed volume's editor and me so that we might pick his brains on the practicalities of conceiving and creating an Oxford Companion. We went to see him at his flat in London and the session was enormously useful. I remarked on the paintings hanging on Osborne's walls, which looked like the works of students, just as we were about to go out for lunch. Osborne confirmed that he had indeed bought them at the local art college, saying that he insisted on being surrounded only by originals, and these were the only kind he could afford. He said he firmly drew the line at displaying reproductions because no printing process can be entirely faithful: the original work's colours can never be reproduced with complete accuracy. The tale he then related to illustrate the point and the feelings it engendered have stayed in my memory. A senior marketing manager from the OUP's New York office had recently visited him in order to discuss plans for the forthcoming *Oxford Companion to the Decorative Arts*, edited by Osborne and due to be published the following October. Over lunch the American had told Osborne that he had a passion for Vermeer's paintings, but had admired them only in books. Oh, said Osborne, there are a couple in the National Gallery: we can go and see them after lunch. So Osborne was present when the American stood before an original Vermeer for the first time in his life. Turning to Osborne after his initial gaze he said, simply, and in a bemused tone: '*It's the wrong green.*'

Entries for the *Companion to Science* were commissioned, written, and edited, but progress was slower than originally expected and it was a long way from completion when I left the OUP at the beginning of 1979. It was eventually abandoned many years later and I have occasionally wondered what might have happened had I stayed. Was the book, embracing the whole of science, too ambitious to work in practice? It was certainly a complicated project and on that account there is unlikely to be a single reason to explain why it wasn't completed. One of the reasons, I suspect, looking back more than three decades later, was that I didn't sufficiently heed Sisam's advice when it came to staving off well-meant suggestions, most especially when their incorporation resulted in a surfeit of detail. Keeping things simple is often an excellent policy.

The *Companion to Science*, approved by the Delegates at the beginning of 1975, had a planned extent of a million words, the same length as Harvey's original volume, and by the time of its acceptance I'd begun to wonder whether there might be room for a modest number of shorter Companions, each devoted to a specific area of science. My colleague John Gillman, biological sciences editor, and I were chatting informally about our publishing interests, as we regularly did, and John mentioned in passing that perhaps there should be a *Companion to Psychology*. I can remember instantly dismissing this as too boring and academic for a general readership. But it must have lodged somewhere in my subconscious and suddenly, I can't recall whether it was days or even weeks later, a title popped unbidden into my head: *The Oxford Companion to the Mind*. Instantly, it took on a life of its own, becoming a sort of catalyst with the potential to generate a book. I had no idea what should be in such a book (I didn't have the specialist knowledge to be able to come up with a list of typical headwords for the envisaged A–Z format of the book) but felt instinctively that, whatever it turned out to be, it would be a winner. Later, when the idea turned into a developing proposal, with a captivating plan and specimen entries, and eventually when I read contributors' articles, my interest in the book grew, as did my conviction that it would be unusually successful, but the original eureka moment sprang purely from the title itself.

I did have a clear idea of who I most wanted to be involved as the book's editor, though: Richard Gregory. Richard was Professor of Neuropsychology and Director of the Brain and Perception Laboratory at the University of Bristol, and the author of the phenomenally successful book, *Eye and Brain* (first published in 1966, its fifth edition finally went out of print over forty years later). He had come to the OUP just over a year earlier with a proposal for a *Dictionary of Scientific Concepts*, which was when I had first met him. The project had by then been going for several years, in that time becoming ever more complicated, with an elaborate superstructure involving many editors, advisers, and consultants, and we judged that it couldn't be made to work and so it was eventually turned down. But Richard had greatly impressed me and he now seemed exactly right for the proposed *Companion to the Mind*. I thought the best way to persuade him would be to visit him in Bristol, taking with me an outline plan of what the book might cover. The wife of one of my authors in Oxford was a clinical psychiatrist and I asked her if she might jot down her ideas on what might be included in the book: this was ready at the beginning of the summer of 1975. I sent a copy to John Lord, the OUP's trade marketing manager, and we met to discuss the project. John said he was unhappy with the proposed title, suggesting that *Oxford Companion to Behavioural Science* or *Human Behaviour* would be more accurate; *Mind*, he said, sounded too academic. It was one of the very few occasions where I thought John was wrong, saying that I took exactly the opposite point of view: I was convinced that the crisper *Mind* was more inviting and less off-putting than John's academic-sounding titles. John was happy to go along and gave me his sales forecast, a suggested initial printing of 20,000 copies and an ideal selling price of £6.50, in terms of values at that time. This enabled me to work out a projected royalty income for the editor, something I hoped would help to tempt Richard Gregory.

I had lunch with Richard in Bristol at the end of June 1975, showed him the tentative outline listing possible topics for an *Oxford Companion to the Mind*, and then asked him if he would agree to be its editor. He said he was flattered, delighted, and full of enthusiasm; and he was keen to move quickly, explaining that he would be leaving for Stanford for a six-month visit in September and would like to be in a position to sign some people up whilst he was there. This meant having a formal proposal ready for the Delegates at their mid-vacation meeting on 15 July. Could it be done in time? Richard agreed to work flat-out preparing the required material, managing to get it to me by the deadline, and the Delegates duly approved the proposal.

The British Association for the Advancement of Science's week-long annual meeting that year was at the University of Surrey in Guildford, starting on Wednesday, 27 August. Richard Gregory was very much in evidence there, as that year's president of the psychology section, and we had a session on the Friday to discuss the book. One thing I was keen to get my hands on was a list of headwords and, as a start, asked Richard to produce one for the letters from A to G. The next time I saw him was on Monday morning in the refectory where breakfast was served, just after 7.30 a.m. Richard hailed and came over to join me. Bubbling over with enthusiasm he told me, as we tucked into our bacon and eggs, that he'd spent ten hours on Sunday generating the alphabetical list of headwords and had completed it, all the way to Z! We had an animated conversation discussing the highlights of the list: it was invariably fun, and exhausting, having a talk with the irrepressible Richard. In a quiet moment the next day I went through the list of proposed entries and in a memo written to John Lord when I got back to the office noted that 'I must say they look good—so many of the headwords make you want to turn right away to the entry and read it there and then.' I ended my note: 'I like the way this one is going.'

Richard was always amusing, and an incident he related from the Guildford meeting has stuck in my mind. It took place at the civic reception that by custom followed the BA's formal opening ceremony. As president of the psychology section at the meeting, Richard had on his lapel the badge worn by each of the meeting's section presidents: a substantial oblong device of metal hanging from its fastening. Richard's group at the time of the episode included the Mayor of Guildford and as Richard bent forward to grasp a canapé from a tray being circulated, his president's badge became entangled with the mayoral chain of office. 'A clash of symbols?' Richard playfully enquired of the Mayor. I am not sure whether or not this elicited a response, but a disentanglement was successfully carried out.

I met a fair number of contributors to the Companion over the next few years but one encounter stands out particularly in my memory. Richard had told me that he was keen to commission an entry on Sufism and that the one person supremely qualified to write it was Idries Shah. A lunch at the Athenaeum had been arranged to discuss the matter: would I join them? After lunch, in December 1976, we repaired to the Athenaeum's first-floor room with its leather sofas and armchairs and here Idries Shah, our host, lit his large cigar. The conversation over lunch had gone well but there was one aspect concerning the proposed entry about which I felt some awkwardness: the fee. The OUP was not noted for the lavishness of its payments for contributors to its reference books, at least not in those days, and for the *Companion to the Mind* we were offering a very modest £20 per thousand words. I raised the matter as delicately as I could, not wishing for there to be any misunderstanding at a later stage, and said I was sorry we weren't able to be more generous. Idries Shah fixed me with the stare of one who has been slighted. I am not writing this article for the money, he declared. I am writing it for Professor Richard Gregory and the Oxford University Press. This is for me a great honour. The episode came back to me just over a year later when I received a splendid letter from Idries Shah's personal assistant, acknowledging a letter I had sent when the entry on Sufism had been received, approved by Richard, and for which payment was being organized. 'Sir,' it began. 'I have the honour to say that your kind letter of 26 January 1978, addressed to The Sayed Idries Shah, has been received. Although unable to answer personally due to his absence abroad, the Sayed has asked me telephonically to express his deep gratitude for your letter and the kindly sentiments expressed in it. In congratulating you upon the progress of the *Oxford Companion to the Mind*, he would like to say how deeply honoured he is to be associated with this project, and that he wishes you all success with it. With assurances of high esteem, Yours faithfully . . .' Those were the days.

I left the OUP a year later but the *Oxford Companion to the Mind* kept going and it was finally published in 1987. Richard's Preface lists the individuals at the Press who took on responsibility for the book after my departure and, with much hard work, a wonderful job they made of it, too. It was my good fortune to be involved at the beginning: a time of excitement, and fun.

The *Oxford Companion to the Mind* went on to sell some 60,000 copies in its hardback and, later, paperback trade edition. Combined book club sales in the UK and USA totalled around 200,000 copies. A second edition of the book, edited by Richard Gregory, was published in 2004 and is still in print.

Adam Hart-Davis left the OUP at the start of 1977 to begin his career in the world of television and I took over his textbooks, including the published titles in the Oxford Physics Series, a series I had started in 1971, around the time of Adam's arrival at the Press. The series had been modelled on the Oxford Chemistry Series, mentioned in Chapter 4 —short, single-topic texts for undergraduate courses—and Adam and I worked on it together. One of the titles in the series going into production at the time of Adam's departure was *Cosmology*, by Michael Rowan-Robinson, then at Queen Mary College in London. I arranged to visit Michael, to meet him for the first time, and over lunch it soon became clear that he loved writing, and what he was keen to do, now that his cosmology textbook was finished, was write a book for the general reader. What would it be about? For the time it took for him to describe what he had in mind, I was transfixed. Suppose you went on a voyage across the universe and your eyes were tuned to see in the ultraviolet, what would you see, and why? Or a journey where you could see infrared radiation, or X-rays and gamma rays, radio or microwave radiation: what would you see, and why? The way he described what would be encountered on such a voyage of the imagination, enigmatic quasars, stars ten times hotter than the sun, stellar graveyards of white dwarfs, dying red giant stars and stars being born, both cloaked in dust, had me completely enthralled. Even the voyage in the familiar visible waveband sounded anything but familiar. The result was *Cosmic Landscape: Voyages Back Along the Photon's Track*, published by the OUP in 1979. A review in *Nature* by D. D. Clayton, an American astrophysicist, recognized just how unusual the book was: 'This is a very good book, although it fails by slim margins to be a great one.' The review was itself highly unusual in that it ended with a poem, 'On His Reading *Cosmic Landscape*', written by the reviewer.

I was never much good at suggesting images suitable for book jackets but in the case of *Cosmic Landscape* I managed to come up with what I thought would be the perfect choice: van Gogh's *Starry Night*. Alas, permission to use it was declined because we weren't permitted to print the book's title over the image. Instead, a detail from a painting by John Martin was used: very fine but not having quite the same impact, in my view, as the van Gogh would have had. A few years later I was in New York and, with an hour or so to spare on a rainy afternoon, I went to the Museum of Modern Art to see, for the first time, the original *Starry Night*. I had been looking at it for a couple of minutes when a young American man standing nearby suddenly addressed me: 'Doesn't it make you wanna weep?' Startled, not used to such directness, I came out with some sort of neutral response, anxious not to get involved in a discussion on art and the emotions. But actually, no, it didn't make me want to weep, though I do remember thinking that the painting was considerably bigger than I'd imagined it would be. And, recalling Harold Osborne's story about Vermeer, I can report that I had no trouble on initially looking at van Gogh's colours, in this particular case his shades of blue.

Planning for the OUP's quincentenary, celebrated in 1978 (a book was first printed in Oxford in 1478, though the origins of the OUP itself strictly date from the middle of the 18th century), began several years before the designated year and editors were asked to give thought to their forward lists: the Press wanted a bumper crop of important and attractive titles to be published in that year, each one to be distinguished by having a specially created 1478/1978 device printed on the spine and title page. Two of my forthcoming trade books were worthy candidates and I duly put them forward for the accolade. One of them was being written by J. Z. Young. I had inherited J. Z. as an author towards the end of 1974 when I took responsibility for John Gillman's biology textbooks on John's departure from the OUP. He was the author of two outstanding textbooks published by the Press, both of which enjoyed remarkably long lives. His *The Life of Vertebrates* was published in 1950 and went through three editions, finally going out of print in 1984; *The Life of Mammals* came out in 1957, with a second edition going out of print at the end of 1996. He retired in 1974 after 29 years as head of the anatomy department at University College London, but was formidably active in research, writing, and lecturing way into retirement. Early on in this active retirement, J. Z. delivered the prestigious Gifford Lectures at the University of Aberdeen during 1975–1977 and these were turned into a book, *Programs of the Brain*, published as a quincentenary title.

The second of my 1978 trade specials had its origins in an editorial trip to Australia in 1973. I had arranged to visit the British scientist Robert Hanbury Brown, head of astronomy at the University of Sydney. Born in India, he was educated in England, trained as an electrical engineer, and was a pioneer of airborne radar in the Second World War, and later of radio astronomy, working with Bernard Lovell at Jodrell Bank. Whilst at Jodrell Bank he invented the intensity interferometer, an entirely new way of measuring the apparent size of the radio sources being studied at Jodrell Bank. Realizing that the technique could be applied to visible stars, Hanbury Brown—the Robert

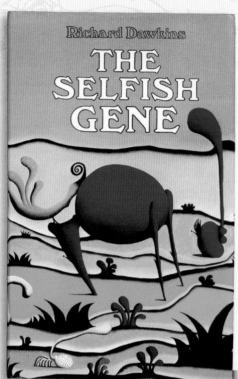

PLATE I
Front panel of the original jacket for
The Selfish Gene, published in 1976:
a detail from *The Expectant Valley*, 1972,
by Desmond Morris. Copyright: the
Artist. 'Eventually we converged on one
painting, with everyone agreeing that
this was the one – its strong central
focus would clearly lend the finished
design a powerful impact.'
(See Chapter 3, page 45.)

PLATE 2
Richard Dawkins, 1980,
in his office at the zoology
department, Oxford.
I was there to collect the
typescript he is holding,
The Extended Phenotype,
published the following year.
(See Chapter 6, pages 92–94.)

PLATE 3

Peter Atkins with *Physical Chemistry* (see Chapter 4) and some of his other textbooks in various editions and translations, taken in 2007. Photo courtesy of Graham Topping.

PLATE 4

Part of the open-plan *Nature* office, then in Little Essex Street, London, 1982. 'There was a real buzz to the place, with the constant sound of typewriters and telephones.'
(See Chapter 6, page 98.)

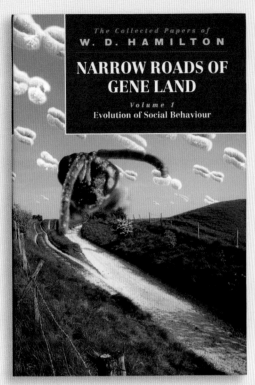

PLATE 5

The idea for the cover illustration of volume 1 of *Narrow Roads of Gene Land* (1996) was entirely Bill Hamilton's: a country lane (taken by Bill himself in his native Kent), a wasp's head (*Nasonia vitripennis*), and cloud-resembling chromosomes. (See Chapter 8, pages 121–125.)

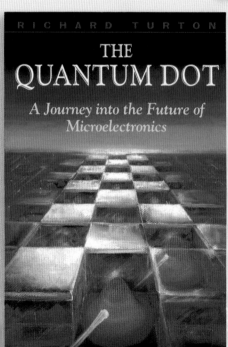

PLATE 6

Pete Russell's design for the cover of Richard Turton's *The Quantum Dot* (1995). Interested readers were directed to a 129-word description, printed at the beginning of the book, of Pete's arresting illustration depicting the atomic world of the quantum dot, the likely basis of electronic devices in the future. It was rejected by the OUP's marketing department in New York for their edition on the grounds of being 'too British'. (See Chapter 8, page 129.)

PLATE 7

The authors of *Organic Chemistry* (2001) during a week devoted to working on the book at a rented holiday house in Herefordshire in the summer of 1997. From the left: Stuart Warren, Peter Wothers, Nick Greeves, and Jonathan Clayden. Another picture of them taken at the same time was used on the back of the cover of the book. (See Chapter 9, page 136.)

PLATE 8

Taken at my retirement party at the OUP, 30 October 2003, this photograph appeared the following week in *The Bookseller* with the following caption: Fond farewell: Michael Rodgers, *third from left*, the OUP editor whose achievements include having commissioned the one million copy bestseller *The Selfish Gene* by Richard Dawkins, retired last week. With him are *from left* his authors John Emsley, Nick Lane, Peter Atkins, Susan Blackmore, Walter Gratzer and Richard Gregory.

had been dropped—searched for clear skies and moved to Australia in the early 1960s, installing the world's first stellar intensity interferometer in the bush at Narrabri, some 350 miles north of Sydney. On the day before my appointment with Hanbury Brown at the university, I discovered by chance a piece of science popularization by him in a Sunday newspaper, one of a regular series of science articles for the lay reader he wrote for the paper, according to the piece. Top drawer scientists with a real interest in popular science were in those days a rarity and I decided on a change to the main topic I had planned earlier for the following day's meeting. The eventual result was Hanbury Brown's *Man and the Stars*, published in 1978.

Over the years I have enjoyed reading many a preface, but have a clear memory of the message of only one of them, Hanbury Brown's. In it we are told that he and his staff at Narrabri Observatory played host to thousands of tourists, most of them having travelled on hot, dusty roads to find out what went on in an observatory. Questions were asked and answered, but

> . . . there was one particular question which always worried me because I couldn't answer it satisfactorily in the time available, usually only a few minutes before the coach left. It was often asked tentatively, away from the main group of visitors, as though it might be thought silly or, perhaps, impolite. It was always the same basic question—'This work may be competent and interesting, but is it any use?'
>
> I found it comparatively easy, given a few minutes, to persuade people that our work was relevant to the general study of astronomy, but I have never been able to abbreviate, to my satisfaction, the arguments that astronomy is an integral part of science and that science is an essential part of our civilization. And so, as the coaches vanished in a cloud of red dust, I was often left wondering what the visitors really thought of our observatory. Many of them, I suspect, felt much the same as they would have done after a tour of a monastery—interesting maybe, but what a useless way to spend one's life!

There was never enough time to talk to the departing tourists about the relevance of astronomy in the broadest sense. The book, the

Preface concluded, contained what Hanbury Brown would have liked to have said, before the coach left.

The British Association annual meeting in 1977 was held at the University of Aston, Birmingham. As usual, I looked through the programme to decide which talks to attend and something must have struck me about what had been planned for the agriculture section: unusual for me because, apart from being a life-long fan of *The Archers*, the BBC radio soap opera, I had no special interest in the subject. The section's president that year was Professor J. K. A. Bleasdale, Director of the National Vegetable Research Station, the NVRS, at Wellesbourne in Warwickshire, and his presidential lecture was entitled 'Britain's green revolution'. As the programme's blurb had it: 'The third world's green revolution is matched by a revolution in Britain's crop production. Science is leading farming through this quiet and continuing revolution. What does the future hold?' I sat in on this session, and another later in the week on 'Science in the vegetable garden', and wondered if there might be a book here. I had a word with Peter Salter, head of plant physiology at the NVRS, who gave the talk after Bleasdale's address, floated the idea of a possible book and arranged to visit Wellesbourne a couple of weeks later.

It is amazing that I was allowed to get away with being involved with a proposal for a book on popular gardening but I gave it the working title 'Science in the vegetable garden' and no one questioned it. John Bleasdale and Peter Salter planned the book, a short paperback to be written for the trade market, and John came up with exactly the right title for it: *Know & Grow Vegetables*. The opening paragraph of the eventual blurb printed on the cover said it all: 'Gardening is usually considered to be an art but in fact "green fingers" are *not* required for producing good vegetable crops. The simple instructions and practical advice given in this book are based on modern scientific research and will work in your garden, because the essential feature of science is that it must be repeatable.' The practical, science-based advice really appealed to me because the one thing I definitely don't possess in the garden is green fingers.

The sales reps loved the book when I presented it at the sales conference just before Christmas in 1978: it was certainly different from the usual fare coming from the science department. The print run was markedly different from the usual, too: 30,000 copies, which were all subsequently sold.

So the quincentenary year of 1978 began well for the College, Reference, and Popular Science department. Peter Atkins's *Physical Chemistry* was published at the beginning of January, the success of *The Selfish Gene* continued to reverberate, two science *Companions* were forging ahead, and that most unlikely of forthcoming science titles, *Know & Grow Vegetables*, was signed up and the writing was on schedule. It felt like a Golden Era, and I thought it would go on forever. But I was wrong: nothing goes on forever. Dan Davin retired later in the year and Robin Denniston was appointed as his successor. I got on well with Robin but it was soon made clear that a structure involving a number of small departments, each separately reporting directly to the head of the division, made no sense. Robin announced his intention to reunite science and to appoint Richard Charkin to run the new department. And Richard made it clear to me, in as friendly a way as possible, that his publishing priority would be centred on specialist monographs, with a back seat for textbooks and popular science. It would mean for me a loss of independence and having to work on the sort of books I no longer found appealing. It was time to move on.

W. H. Freeman's UK office had decided to expand their editorial department and approached me about joining them. The timing was right and so at the beginning of 1979 I left the OUP.

r- AND *K*-SELECTION
AND THE
EXTENDED PHENOTYPE

W. H. Freeman and Company was founded in 1946 in San Francisco. The first book it published, a general chemistry textbook, was by the renowned Linus Pauling; and later celebrated texts included Lubert Stryer's *Biochemistry* and Jim Watson's *Recombinant DNA*. I have a tee shirt, produced in the 1990s by the marketing department for a sales conference, listing no fewer than ten Nobel laureates published by Freeman: an astonishing number for any publisher, let alone one of Freeman's modest size. The firm was also noted for the outstanding design and high production standards of its books and catalogues. I had visited them in San Francisco towards the end of 1976, taking for signature the OUP contract for the North American rights for Atkins's *Physical Chemistry*, and enjoyed meeting its enthusiastic editors, appreciating the lively atmosphere they created. So, at the beginning of February in 1979 I reckoned this was a good place to be for the next stage of my career as a commissioning editor.

Freeman's UK office, originally set up to market and sell the parent company's books in the UK and continental Europe, was then in Reading. It was a small operation—16 on the staff was the maximum during my time there—and the accommodation was cramped and basic. John Gillman, an editorial colleague at the OUP at the beginning of the 1970s, had joined Freeman at the end of 1974 to build a UK-based list of books but our period back together again proved to be exceptionally short, three weeks, ending with John moving on to another publisher. In charge of Freeman's UK outpost—its managing director—was Jonathan Backhouse, Sir Jonathan Backhouse, Bt, to give him his formal title. Jonathan could be charming, though he had an acerbic side too: negotiating with him was sometimes agreeable and amusing, on other occasions difficult, even intimidating.

Encouraged by the senior management in San Francisco, the decision was taken not long after my arrival to relocate Freeman's UK office from Reading to Oxford and this was finally achieved during that first summer, in 1979, when we moved to stylish premises in Beaumont Street ('the finest street ensemble of Oxford', as Nikolaus Pevsner describes it in his *Buildings of England*). At the end of our first day there only Jonathan and I remained, our colleagues having departed for home early. We were in Jonathan's elegant office with its chandelier, cherrywood desk topped with red leather, an oil painting waiting to be hung, and several unopened packing cases. A bottle of superior claret, bought by Jonathan to mark the occasion, was produced and opened. Our moods were very different. I was delighted to be back in Oxford where numerous friends, former colleagues, and authors were located. Jonathan was afflicted with a sense of foreboding, citing examples, real or imagined, of the fate of organizations that, doing well whilst working in modest surroundings, eventually take the fateful decision to move to grander ones. I'm not sure how successful I was at dispelling his gloom; I did my best, but we certainly enjoyed the claret.

In September I flew to San Francisco, keen to bond with my new colleagues, as well as do some campus visiting. W. H. Freeman is owned by *Scientific American* magazine and each year a board meeting was held in San Francisco, attended by the president and directors of Freeman. Afterwards the editors joined for lunch, which, along with the earlier board meeting, took place in the Olympic Club. We were not long into lunch when the president approached me and whispered in my ear. Apologizing for putting me on the spot without prior notice, he said that the president of *Scientific American* and his board were curious about the newly appointed editor from England. Would I make a little speech . . . *now?*

In my prologue I mentioned that whenever I learned something new, a grand idea or a small but pleasing detail about science, my automatic reaction was to want to relate it to others. This desire to communicate appealing ideas came to my rescue that lunchtime, saving me from being rendered speechless, with nerves making my mind go blank, when a spoon hammered on a table had brought about an instant silence and I rose to my feet. The small but pleasing detail about science that flashed into my mind as a suitable basis for my speech was, of all things, the r- and K-selection concept of population biologists. I had read about it in *New Scientist* and what had made it memorable was that the writer had waggishly suggested a perhaps illuminating, certainly amusing, parallel with book publishers. The r and K, symbols in the algebra used by population biologists, refer to the qualities needed by breeding organisms to succeed in different sorts of environment. If an environment is stable and predictable, there will be intense competition between individuals for the limited available resources. In such an environment it pays to have only a small number of offspring, and to care for them for a relatively long period, after which they will be well equipped to survive on their own. This is the K-strategy and the other qualities selected include large size and long life. I invited my audience to think of an elephant as an example.

If on the other hand the environment is unstable and unpredictable, there is no point in caring intensively for a small number of offspring: it pays instead to do the reverse and produce large numbers of small offspring, leaving them to their own devices. Statistically, a proportion will survive. This is the r-strategy and there are examples from both plants, such as weeds, and animals. It wasn't difficult to carry my audience with me when I spoke of 'r publishers', churning out large numbers of specialist tomes, and 'K publishers', concentrating on a limited number of textbooks, or trade books, and lavishing on them substantial editorial and marketing resources with the aim of succeeding in highly competitive but potentially large markets. W. H. Freeman, I pointed out, was famously a K publisher. So I was pleased that I had joined them because I felt that I was a K editor.

The speech went down well: I had been rescued by the concept of r and K. It was now time to return home and put the theory into practice. Which 'K books' to pursue?

On the morning of Sunday, 7 May 1978, when I was still at the OUP, Richard Dawkins and I went to church together. As atheists, we were not especially interested in spiritual matters but we did want to hear Arthur Peacocke's Fifth Bampton Lecture that morning in the University Church of St Mary the Virgin. The Bampton Lectures, established in the 18th century, comprise eight divinity lectures delivered annually in Oxford. Arthur Peacocke was an ordained Oxford biochemist and the title of his fifth lecture, *Creation and the world of science: 'The selfish gene' and 'What men live by'*, had caught Richard's and my interest.

I am not able to remember anything of the lecture but the day was memorable because, on leaving the church, Richard invited me round to his room in New College to talk about the book he wanted to write next, his follow up to *The Selfish Gene*. My earlier crash course in biology whilst working on *The Selfish Gene* had resulted in my understanding of the difference between an organism's *genotype*, its full set of genes, its genetic makeup, and its *phenotype*, meaning its outward characteristics, such as the colour of an eye, or of the hair of an animal, or of the flower of a plant. I had also taken in the fact that not all phenotypic characters can be seen: some are invisible, residing for example in the organism's biochemistry. And I had remembered that a phenotype is not determined solely by the genotype: it is also influenced by the environment. So although an individual's height potential is genetically determined, the actual adult height reached depends on external factors, such as the quality of nutrition in childhood.

What Richard introduced me to, that Sunday morning, was the idea that one can think of the influence of genes inside a body extending beyond the body (its phenotype) and that gene differences will result in differences in that influence. This next book was to be *The Extended Phenotype* in which, in words used by Richard in his eventual Preface, he '. . . attempts to free the selfish gene from the individual organism which has been its conceptual prison'. The next sentence is equally dramatic: 'The phenotypic effects of a gene are the tools by which it levers itself into the next generation, and these tools may "extend" far outside the body in which the gene sits, even reaching deep into the nervous systems of other organisms.'

Richard described an example illustrating this unsettling notion, continuing with his outline of the book he had in mind. For me, this specific story belongs firmly in the 'once heard, never forgotten' category. In the book itself, Richard discusses how parasites manipulate the behaviour of their hosts, going on to quote from N. A. Croll, the author of a publication from 1966. In this specific case the larvae of a particular parasitic worm begin life inside an insect but need to break out of the host and get into water where they live as adults. The way the parasite brings this about is so astonishing that I reproduce the passage here:

> . . . a major difficulty in the parasite's life is the return to water. It is, therefore, of particular interest that the parasite appears to affect the behaviour of its host, and 'encourages' it to return to water. The mechanism by which this is achieved is obscure, but there are sufficient isolated reports to certify that the parasite does influence its host, and often suicidally for the host . . . One of the more dramatic reports describes an infected bee flying over a pool and, when about six feet over it, diving straight into the water. Immediately on impact the gordian worm burst out and swam into the water, the maimed bee being left to die.

Richard wasn't ready to prepare a formal proposal until the following year, by which time I was with W. H. Freeman and I signed the book up for them. *The Extended Phenotype* was being written for a professional readership but I read each completed chapter with enjoyment. As with *The Selfish Gene*, the writing, so fluent, carried me along; the arguments were complicated but beautifully explained; and along the way were wonderful stories from animal behaviour. In his Preface, Richard stated that while the book assumed the reader had professional knowledge of evolutionary biology and its technical terms, '. . . it is possible to enjoy a professional book as a spectator, even if not a participant in the profession'. That is exactly right, and in the few cases where an author is able to pull it off, lay readers experience exhilaration, and a sense of privilege, coming from a vicarious feeling of being at the frontiers of science.

The publication year of *The Extended Phenotype* as printed in the book is 1982, but the actual publication date was 14 December 1981. A small reception at the W. H. Freeman office in Beaumont Street was held for Richard at the end of that afternoon. For the office staff it was the first time that they'd seen an author in the flesh and on that account it was a special occasion, giving me particular pleasure.

Paperback rights for *The Extended Phenotype* were eventually sold to the OUP. At the time of writing (2013), this edition was still selling, more than thirty years after the original publication (see Plate 2).

Another unorthodox book of that time started when I met Peter Atkins for a drink we'd arranged to have sometime in 1978, after his *Physical Chemistry* textbook had been published. In the middle of our chat, out of the blue, he began to talk about a lecture he was to give during a forthcoming visit to America. He said he wanted to pose the most basic questions about our world, our universe. Why are there only three dimensions of space? Why does time run forwards? Intrigued, I asked if it was possible to provide accessible answers to such questions. The answer was yes, and I was hooked. Peter later gave the lecture but the ideas continued to evolve. Suppose you wanted to create the universe, what would you need to specify? Everything? Or might it be possible to stipulate a limited number of entities from which, inevitably, everything else would follow? The question would be of interest to a lazy creator.

Going further, what would you need to design if you were an *infinitely* lazy creator? What would be the minimum you could get away with, after which everything would automatically follow? Peter wanted to persuade readers that the universe could come into being without the need for a creator, along the way explaining why the universe is the way it is, and culminating in a description of how it came into being. This was a big canvas, but the book was to be unusually short. Its purpose was spelt out in the wonderful opening paragraph:

> I shall take your mind on a journey. It is a journey of comprehension, taking us to the edge of space, time, and understanding. On it I shall argue that there is nothing that cannot be explained, and that everything is extraordinarily simple.

The book was *The Creation*, published in 1981, and there were strong reactions from both ends of the spectrum. Critics ranged from the author of a negative review in *Nature* to an irate vicar who returned his copy of the book and demanded his money back. At the other end of the scale, one reviewer described the book as 'brilliant and challenging', and another as 'exciting and memorable'. That last quote provides a fair description of what it felt like working with Peter, from first hearing about the book to delivery of the final typescript some three years later.

Two other books commissioned and published by W. H. Freeman during the early 1980s were unusual in different ways. One was *The Evolution of Love* by Sydney Mellen, published in 1981. It had seemed an unlikely prospect when, having virtually completed the book, Mellen contacted me. A retired American diplomat in his early 70s, his exploration of the biological origins of love—how the emotion developed in our ancestors, how natural selection favoured various kinds of love more strongly in humans than in any other species—covered fields ranging from zoology, palaeontology, and archaeology, to genetics, ethnology, and psychology. All very interesting, but could such a broad-ranging foray into so many fields, and by an amateur, really be brought off? It could, with the typescript surviving and benefiting from a rigorous process of reviewing by numerous academic specialists. One academic who enjoyed reading the typescript was the distinguished South African archaeologist and specialist on the early prehistory of Africa, Glynn Isaac. At the time Professor of Anthropology at the University of California at Berkeley, Isaac gave the book his seal of academic approval by contributing a Foreword.

The second unusual book was David Jones's *The Inventions of Daedalus: A Compendium of Plausible Schemes* (1982). David was the author of the Daedalus column which appeared weekly in *New Scientist*, and had started in 1964. Each of his pieces, a flight of inventive fantasy but always starting from scientific reality, proposed an invention, ingenious and novel, that made the reader not only smile but also think. Might it actually be possible, or was it wholly fantastic? The answer was rarely obvious. At the time I got to know David, when we discussed the possibility of basing a book on a selection of his pieces (in these conversations he disconcertingly referred to himself in the third person, as Daedalus), around 20 per cent of his proposed inventions made it into the real world: they worked! It was fun collaborating with David, and I discovered along the way an amazing range of useful facts, from the military advantages accruing from wearing non-Newtonian trousers to recovering the long-lost work songs of ancient Greek plasterers.

Working on occasional unconventional books, whether on the creation of the universe, the evolution of love, or the inventions of Daedalus, was absorbing, and certainly satisfying when they went on to sell well. Being able to pursue them was possible because of the editorial freedom I was given—an important, and pretty unusual privilege. That said, a science publishing programme, if it is to have a long-term future, also needs an underpinning of exceptional mainstream titles: 'K books' for student and professional markets. Two in particular in this category gave me a real sense of achievement when I commissioned them: I felt certain that when they were published they would stand out from the crowd, and go on selling over the years. Tom (Thomas E.) Creighton's *Proteins: Structures and Molecular Properties* and Michael Rowan-Robinson's *The Cosmological Distance Ladder* went on to do just that, becoming classics in their respective fields.

I should say something about the above-mentioned unusual privilege of enjoying so much editorial freedom, not least because it will seem strange to readers today who are in the early stages of commissioning editor careers. As I mentioned at the beginning of this chapter, the W. H. Freeman office in the UK was small. Its principal function was to market and sell the American parent company's books in the UK and continental Europe, and everyone in the UK office was engaged on doing just that. Everyone, that is, except me. I was the only one there with direct experience of commissioning, and that was why I had the freedom to pursue books and authors of my choosing.

Dramatic news from W. H. Freeman's headquarters in San Francisco reached the UK office in the middle of February 1980. The head of Scientific American had, without warning, descended on the Freeman office and invited the president to clear his desk within the hour. We were told that the new president, Neil Patterson, had a track record as a brilliant editor, and that he was exactly the man to transform W. H. Freeman from its torpor to a dynamic presence in the publishing firmament. Yet in Oxford we were alarmed rather than inspired because Neil's first action was to fire half of the editors in San Francisco: the St Valentine's Day Massacre seemed to us in the UK to be a fair description of this cull. But I was not to worry: Neil, I was told, admired my record as an editor and I was safe. This reassurance was repeated in person when Neil visited the UK a couple of months later.

Gradually, though, during the following year, uncertainty over the future of Freeman's UK editorial programme grew. Neil was channelling considerable resources into a new Scientific American book club and Oxford's publishing was still not mature enough to be making money. During a visit to Oxford in the summer of 1981, Neil expressed to me his worry that an editor working from the UK might be condemned, because of the size of the country, to a perpetual 'frenetic scratching around'. But I was given the chance to prove him wrong and was asked to prepare a report on my publishing programme. Jonathan Backhouse subsequently wrote to Neil with a financial projection: its conclusion was that, provided my authors delivered their completed manuscripts by the promised dates, 'it appears that we are likely to turn the corner in the current year and will be substantially profitable by 1983'. Jonathan recommended acceptance of my request for a year's grace, followed by a similar exercise during the following summer.

A period of grace was granted although, as it turned out, this didn't last for a year. In March 1982, Jonathan told me that the decision had been taken to call it a day, the reason being that the promised delivery dates from some of my authors had slipped, meaning that the previous summer's financial projection no longer held. The End, then, though I was given plenty of time to clear my desk, finally leaving a couple of months later.

The campaign to find another job began shortly after receiving my bad news in March but the various leads I followed up, each one seeming promising at the time, came to nothing. At the beginning of July, a friend showed me an ad in the journal *Nature*, inviting applications for the temporary post of locum physical sciences editor. In those days the *Nature* staff had a three-month period of sabbatical leave every three years and the advertised job was to cover the absence that summer of the physical sciences editor, Phil Campbell. There were two problems. I wasn't qualified (the ideal candidate described in the ad was 'an academic probably doing research in astronomy'), and the closing date for applications had long since passed. I consulted another friend and he advised me not to be put off by such minor problems. He knew John Maddox, editor of *Nature*, and said he would ring him forthwith to find out if I might be in with a chance. Within minutes I was telephoned by Maddox and interviewed. At the end of our conversation he said that if I could come to the *Nature* office that afternoon, Friday 2 July, he would arrange for Phil Campbell to interview me. Phil's interview included a test: he left me to study a research paper recently submitted to *Nature* and then quizzed me on it. As soon as this was over, I was summoned to meet John Maddox, who simply said: 'We'd like you to start on Tuesday.'

I was given a temporary contract, lasting until the end of October, when Phil would be returning from his sabbatical leave. My job was to take responsibility for the assessment and final recommendation on the publication of research papers, mainly on astrophysics but also on cosmology and physics. Unsolicited manuscripts arrived each day to be read and either rejected there and then or sent out for review by specialists. The *Nature* office, then in Little Essex Street, just off the Strand, was open plan, and in one corner was John Maddox's separate, glass-panelled office. There was a real buzz to the place, with the constant sound of typewriters and telephones (see Plate 4).

Nature was, and is, a curious mix of research papers, where there might typically be a modest number of months between submission and eventual publication, and the magazine part of the journal: the front, with its news stories and—in those days—John Maddox's editorials, these elements and others governed by the harsh deadlines of a weekly publication. The demands of the weekly cycle meant there was always a sense of urgency, which is why I found it such an exciting place.

I was introduced to the weekly News and Views meetings, held each Tuesday morning in John Maddox's office. The News and Views column, in the journal's magazine part, provided commentaries, accessible to a broad readership, on a few selected papers published in that week's issue. At the meeting, editors put forward any paper, accepted during the past week, which might be suitable for this treatment. This meant that aspects of the research might interest *Nature*'s general readership, if skilfully opened out and explained in a specially commissioned piece. At the end of the meeting—invariably a challenging occasion, with John Maddox asking probing questions—the News and Views editor would return to his desk and attempt to commission such pieces. These meetings brought home to me the vastness of my ignorance of large areas of current biology. A few years later, I commissioned Eleanor Lawrence's *A Guide to Modern Biology*, a book for students published in 1989, and the motivation for doing so can be traced directly back to those meetings, and my realization of how little I knew about what was happening in important and exciting areas of science. And there was another, pleasing link with the Tuesday meetings. Eleanor had been News and Views editor during the late-1970s.

My campaign to find a permanent job continued over the summer and, with the blessing of *Nature*, I occasionally popped out to attend interviews. On 14 October, with two weeks to go before my contract with *Nature* expired, I was invited for an interview with the publisher Longman in Harlow, Essex. Robert Welham, managing director of Longman's university division, rang me shortly after I had got back to my desk at *Nature*, offering me the job of running and growing Longman's tertiary science publishing. I began doing that at the start of November 1982.

CHAPTER SEVEN

THE BLIND WATCHMAKER
AND THE
UNIVERSE IN
TWENTY OBJECTS

AFTER THE UPHEAVALS of my final months at the OUP, followed by the uncertainties hanging over my survival during the unsettling last year at W. H. Freeman, I was more than ready for a stable environment when I arrived at Longman on 4 November 1982. Everything certainly felt exactly right, this initial impression reinforced by the warm welcome I received from friendly colleagues.

Founded in London in 1724, Longman was the oldest commercial publishing house in the UK. Most of its publishing moved out of London at the end of the 1960s to Harlow, into premises specially designed by Frederick Gibberd, the architect appointed to plan Harlow New Town, inaugurated in 1947. Longman's publishing at the time of my arrival ranged from textbooks for schools in Africa and English Language Teaching materials, the two biggest of the seven publishing divisions, to the smallest, the tertiary and further education division, which I was joining. The science list was modest, accounting for less than a fifth of the division's turnover, but that was fine: I liked the idea of growing new things from small beginnings.

Robert Welham gave me a few weeks to settle in and then casually mentioned that I featured in the Longman Board's latest Strategic Development Plan. Well, it was in the form of a footnote of a dozen or so words, stating that I was to submit a plan for the future development of tertiary science publishing. The experience of carrying out this instruction proved to be epiphanic, which is why I remember it. I had been commissioning books for ten or so years yet I suddenly realized, writing my Plan, that all of my basic assumptions needed to be revisited. Why concentrate on these particular areas? Why focus on certain kinds of book? The answers were not as obvious as I'd been supposing. I saw for the first time how it was all too easy, when enthusiastically building a list of books, to begin to take for granted, and then forget about, the assumptions underlying a commissioning strategy. It was all very salutary and I became an evangelist for Strategic Plans, believing that *everyone* should regularly be asked to work on one. Colleagues at lunchtime humoured me as I attempted to persuade them, but I ended up convincing none of them.

I am not able to remember the detail of what I set down in my Plan, but it was broadly concerned with characterizing the different markets in science, from undergraduate courses to professional reference, from biology to physics, and not least, to popular science. What were the growth areas, in terms of student numbers or research funding, and which of them were contracting? I do remember feeling some irritation at the outset for having to invest so much time in the exercise but once my Plan had been submitted, and Robert Welham and I met Tim Rix, the Chief Executive, to discuss it, I did feel better for having had to do it. I suspect that the vast majority of academic commissioning editors in science at that time were so busy with their daily workload that they simply didn't have the time to think enough about and analyse their commissioning assumptions.

The basic problem with Longman's science publishing was that it had been drifting for the previous ten or so years, with an emphasis on traditional, unexciting areas. It was associated with dullness and the old-fashioned, and sales of too many individual titles were disappointing. It was going to take many books and several years to turn things around but there was one specific book in my mind that encapsulated everything I wanted to achieve: a high-profile title with the potential to become a bestseller. This was Richard Dawkins's next book, *The Blind Watchmaker*, and I had known about it before my arrival at Longman.

Longman in Harlow hadn't published a major trade book like *The Blind Watchmaker* before, certainly not on the science side, and an early task was to win over my sales and marketing colleagues, persuading them that we could handle the new Dawkins and make a success of it. The next challenge was to convince a sceptical Caroline Dawnay, now Richard's literary agent, that Longman was up to the job. Longman, Caroline had correctly pointed out when I first contacted her from there (I had worked with her earlier when signing up Richard's *The Extended Phenotype* at W. H. Freeman), had no track record of selling a book like this, but she agreed to meet for lunch and listen to my case. This was at the beginning of May 1984 and at the end of lunch Caroline delivered her verdict: Longman would be allowed to enter the competition to win the book. Rival publishing houses were to be invited to bid in an auction at the end of the summer.

One of Caroline's requirements was that each bid should combine hardback and subsequent paperback rights. I saw this as helpful for the Longman cause because a sister company was Penguin—both publishers were owned by Pearson—and I felt there could be no better partner in the campaign to win the book. I contacted Penguin's science editor, Donald McFarland, and we met for lunch in late July. Donald enjoyed hearing about the book and confirmed that Penguin would be delighted to be involved. We agreed to talk to our respective chiefs about how much could be offered to make up the combined royalty advance. Money wasn't going to be everything, at least I hoped that was the case, and I was keen to set down exactly how Longman would market and sell the book. The result was a detailed marketing plan, written in collaboration with my trade marketing and sales colleagues. It was sent to Caroline before the deadline for bids expired, together with the offered Longman / Penguin advance.

The auction itself, the first I'd taken part in, was in early September, and was a nerve-jangling experience lasting four hours. Caroline rang me a few seconds—literally—after the appointed starting time, noon. She asked me to look again at various minor details of our offered terms, with a view to improving them, and finally, no surprise, added that we would need to come up with more money. My boss at that time, Robert Duncan, was in Edinburgh that day and I rang him there. He was nervous about offering more money. Was the book worth it? Please would I convince him. There was eventual agreement to increase Longman's share. Then a telephone call to Donald McFarland at Penguin. He would have to talk to his boss and get back to me. The message when he did call back was yes, they could come up with more, but this represented their absolute limit. I phoned Caroline with our revised offer and waited for her final call. This came at last: Longman had won and secured the book. 'Michael,' said Caroline, 'you are the luckiest publisher in all London!' I smiled at the dated effusion, sweetly meant, but later wondered if she might be right.

Richard planned to finish writing by early 1986 and he sent a draft of each completed chapter for comments. It was not a repeat of *The Selfish Gene* experience, in that there can be only one first time, one experience of being completely bowled over by something that has never happened before, but that said, it was a wonderfully exhilarating time. Once again, when reading the chapters, I found myself occasionally pausing to allow nervous energy of excitement to dissipate.

At the beginning of August in 1985 my wife and I were invited to Saturday lunch at the home of my Longman colleague Guy Garfit and his wife, Julia. Their other guests were the actress Barbara Flynn and her husband, Jeremy Taylor. Barbara and Julia had known one another since they were at school together; Jeremy was a television documentary-maker at the BBC: Guy and Julia had arranged the lunch so that I could talk about *The Blind Watchmaker*. Jeremy went on to write and produce 'The Blind Watchmaker' for the BBC's *Horizon* series: presented by Richard Dawkins, the programme went out in January 1987, a few months after publication of the book.

The elaborate plans for launching the book required the 1986 publication date to be set during the previous year, and this in turn meant we needed to have at the same time a firm delivery date for the completed typescript. Richard committed himself to a completion date of Sunday, 16 February 1986 and we agreed that I would collect the typescript from him at his home in Oxford on that day, at 10.00 a.m.

The day duly arrived and I can recall the exact time when I walked up to the front door and rang the bell: it was 9.50 a.m. Richard's wife opened the door and greeted me warmly, though her spontaneous words of welcome were: 'You're early!' I was taken into the sitting room and a few minutes later Richard appeared, in a hurry, clutching a few pages that had just spewed out of his printer—I think his Preface. He gave me a quick smile and whilst checking the freshly printed pages uttered the words: 'You're early!'

As with *The Selfish Gene*, I was keen that *The Blind Watchmaker* didn't contain illustrations, feeling that they would be a distraction for the reader, but my sales and marketing colleagues argued that pictures would make the book more attractive and sellable. A compromise was agreed and the artist Liz Pyle was commissioned to create a series of drawings to face the chapter openings, in addition to one serving as a conventional frontispiece, the idea being that they would imaginatively capture the spirit of individual chapters. Initially uneasy, I was won over as soon as I saw Liz Pyle's magnificent drawings, so powerfully atmospheric, and wholly in keeping with the book. We were on the point of asking Liz to take responsibility for an illustration for the jacket when Richard telephoned to tell me that Desmond Morris had been so intrigued by his title, *The Blind Watchmaker*, that he had been inspired to paint a picture with the same title. Would I like to come over to Oxford to see it, with a view to its being used for the jacket? We did use it, as did Penguin for their later paperback edition.

The American rights for *The Blind Watchmaker* had been acquired by the New York publishing house W. W. Norton, and it had been agreed with them that Longman would be responsible for the editing and typesetting. When the time came we would supply Norton with film, from which they would print their edition. The two editions would therefore be expected to be identical, apart from the jackets and the two versions of the title and copyright pages giving details of the respec- tive publishers. One, tiny detail was different, however, and I mention   it because the reason is, if nothing else, amusing. Richard opens one chapter by observing that many people find it hard to believe that something as complicated as the eye could have arisen from small beginnings by a gradual series of step-by-step changes. For them, an eye could be useful only if it were fully formed: intermediate stages would confer no advantage. The answer to the question, 'What use is 5 per cent of an eye?' is that 5 per cent vision is more useful than no vision at all. The argument in the book moves on to animal mimicry as a way of gaining protection from predators: stick insects, for example, look like sticks and so are saved from being eaten by birds. At this point the narrative mentions the Harvard palaeontologist, Stephen J. Gould, and his remark on dung-mimicking insects: 'can there be any edge in looking 5 per cent like a turd?' We now come to the point of this tale, which resides in the book's index. Its rubric begins: 'This book is meant to be read from cover to cover. It is not a work of reference. Many items in the index will mean something only to people that have already read the book and want to find a particular place again.'

Fair enough, and one can imagine a reader being fascinated on first reading about dung-mimicking insects and wanting to locate the passage later. In the index there are seven subheadings under Gould, S. J., one of which reads 'five per cent resemblance to turd'. Norton in New York, publishers of Gould's books, were outraged, convinced that Dawkins was ridiculing their distinguished author, and nothing could persuade them otherwise. That particular subheading stayed in the UK edition but was removed from the film used to print the American one.

The Blind Watchmaker was published at the end of September 1986, and the sales achieved by Longman surprised many, both inside and outside the company. The subscription for the UK (the number of advance orders placed by the trade ahead of publication) totalled more than 17,000 copies: especially impressive when compared with the 18,000 or so copies of the American hardback edition sold by Norton before they published their paperback. Longman went on to sell 25,000 hardbacks before, 18 months later, Penguin's paperback was published.

Involvement with successful trade books is fun and exciting but I also found academic publishing satisfying in different ways, enjoying the mix of the two. I mentioned at the beginning of Chapter 4 that many famous science textbooks have prosaic titles—they are simply the titles of the courses for which they were written—and those of a clutch of especially attractive academic books in the list I inherited at Longman will sound to an outsider prosaic in the extreme: Kaye and Laby's *Tables of Physical and Chemical Constants*, and two textbooks by Arthur I. Vogel: *Practical Organic Chemistry* and *Quantitative Chemical Analysis*.

Attractive? All three books were—still are—well known in universities and laboratories throughout the world, and all three had been selling steadily, making real money, for decades. *Kaye and Laby* first appeared in 1911 and after Kaye's death in 1941 successive editions were prepared under the direction of editorial committees. A fifteenth edition was already in preparation when I arrived and this was published in 1986, celebrating the book's 75th birthday. Vogel was an astonishingly prolific author, writing from the mid-1930s to the mid-1950s a series of textbooks on chemical analysis and on practical organic chemistry, and they continue to stand the test of time. He died in 1966 and new editions of his main texts were then prepared by small teams of authors from the chemistry department he led for over thirty years at a college in Woolwich, in south-east London. I took responsibility for the fifth editions of *Practical Organic Chemistry* and *Quantitative Chemical Analysis*, travelling regularly to Vogel's old department at what later became the Thames Polytechnic in Woolwich, eventually the University of Greenwich, and, in spurring-on mode, joined the respective teams. Both of these new editions were published in 1989.

The Vogel new editions, and the Kaye and Laby, were worth the investment of the time and effort they needed because they were going to generate healthy sales incomes for many years. They were good examples of the *K* book, mentioned early in the previous chapter, and it was satisfying to bring them to a successful conclusion. Another sort of satisfaction on the academic side came from commissioning and publishing two authors I had first met in Oxford whilst working on *The Selfish Gene*. Mark Ridley at that time, the mid-1970s, was a graduate student working with Richard Dawkins and we had kept in touch after I left the OUP. At Longman I commissioned his short, elegantly written text, *Evolution and Classification*, published in 1986.

The second author was Marian Dawkins, Richard's first wife, who, like Richard, was in the zoology department at Oxford. Marian, an exceptionally clear writer and a patient expositor, wanted to write a textbook that would make comprehensible a number of areas in animal behaviour that students regularly found difficult. Her working title, *Stumbling Blocks in Ethology*, precisely spelled out her purpose but it was hopelessly clunky and off-putting, which we both knew at the outset. We were confident that a more appealing title would eventually emerge, but as the writing approached completion inspiration continued to elude us. Marian posted a notice in her department offering a bottle of champagne to any of her colleagues who could come up with a good title. The winner was Richard Dawkins with his suggested *Unravelling Animal Behaviour*, and the book was published in 1986. It is worth taking trouble over a title.

I want briefly to mention here a trade paperback, published in 1985, partly as a reminder of a subject many people found troubling at that time: the possibility that a full-scale nuclear war could lead to a nuclear winter, the result of enough smoke and dust being injected into the atmosphere to blot out sunlight over the entire globe. In the extreme case, some scientists argued that this could result in the extinction of all human life. I first heard about the subject from Michael Rowan-Robinson, then at Queen Mary College, London, where he worked on cosmology and infrared astronomy. Michael and I had collaborated on his *Cosmic Landscape* at the OUP, and *The Cosmological Distance Ladder* at W. H. Freeman. He wanted to write for a lay readership a straightforward account of the nuclear winter effect, its causes and consequences. His motivation came not only from his personal views on nuclear weapons—he supported the work being done by Scientists Against Nuclear Arms (SANA)—but also from his professional studies of stars shrouded in dust, a field centrally relevant in following calculations on a nuclear winter. The result was *Fire and Ice: The Nuclear Winter*. It sold only modestly, a few thousand copies at most, but one sales quantity turned out to be wholly remarkable, and this is the main reason for my mentioning the book. My colleague Lynette Owen, rights director at Longman, had sold the Japanese translation rights for *Fire and Ice*. In hindsight, had I given it more thought at the time, it would probably not have been difficult to guess that there might well be some special interest in the book's subject in Japan. But I don't believe anyone could have foreseen the extent of that interest: the Japanese edition, in print until 1993, sold an astonishing 76,000 copies.

I had lunch with Michael not long after *Fire and Ice* had been published to talk about ideas for his next book: which of them, he asked, should we go for? Finding the most ambitious proposal irresistible, it was for me an easy choice. Michael's idea was to focus on twenty famous astronomical objects spanning the whole range of what there is in the universe, from comets to quasars, a chapter for each, and through them communicate all that is known about the universe. The style would be accessible to readers with little or no scientific background, the narrative enriched by the inclusion of myth and historical record, and poetry and paintings with an astronomical dimension. Several hundred full-colour illustrations would include images from the world's great ground-based and satellite-borne telescopes. In a sentence, the book would survey everything there is in the universe, bringing together history, mythology, literature, art, and the frontiers of modern science. Here is a taste of this bold, all-embracing approach, taken from the chapter devoted to Polaris, the Pole Star, in *Universe*, published in 1990:

> Edmund Spenser, in the *Faerie Queene*, writes both of Polaris and Ursa Major, compressing into a few brilliant lines the relationship of the Plough and the Pole Star, the fact that the Plough never sets, the steadfastness of the Pole Star and its use for navigation:
>
>> By this the northern wagoner had set
>> His sevenfold team behind the steadfast starre
>> That was in ocean waves never yet wet,
>> But firme is fixt, and sendith light from farre
>> To all that in the wide deep wandering arre . . .

One of my ambitions for the popular science books I worked on over the years was that at least a few—I always hoped it would be more than that—would be enjoyed by readers with a non-scientific, humanities background, and I certainly felt this strongly about *Universe*. The book was favourably reviewed, and sold well, but I had no way of knowing how many, if any, readers with an arts background were touched and intrigued by it.

Norton's Star Atlas is the most famous astronomical atlas and reference handbook in the world, first published in 1910. In 1985 *Norton's* was in its seventeenth edition and its owner, a one book, cottage-industry publisher, wishing to retire, invited offers from a number of publishers to acquire it. At that time, when the details landed on my desk, I had never heard of the book, and I knew little of the world of amateur astronomers, but it didn't take long to realize that here was a prize worth going all out to win.

Up until then I had assumed, without really thinking about it, that the positions of the stars were all but constant. No, I discovered that morning in 1985, they slowly change position on account of the earth's precession (the slow wobble of the earth on its axis, like that of a spinning top, caused by the gravitational pulls of the sun and the moon). The star maps in the seventeenth edition of *Norton's*, published in 1978, had originally been drawn to show positions in 1950, Epoch 1950.0 as it is formally designated. A major task for a new edition would be to produce new maps for the year 2000, Epoch 2000.0.

There was competition from other publishers—it was a highly desirable book for those who could see that—and I was determined to pull out all the stops. This involved enthusing the sales department, getting from them the ambitious sales forecast needed to justify a generous royalty advance. My boss approved the proposed advance and I met the book's owner to make the case that we were the right publishers.

We eventually secured *Norton's* and brought the seventeenth edition back into print in 1986. Ian Ridpath, writer and broadcaster on astronomy and editor of *Popular Astronomy*, was appointed editor to take charge of the new edition. *Norton's 2000: Star Atlas and Reference Handbook* was published in 1989. It was a K-book *extraordinaire* and I was able to tell the sales force at the conference when it was presented to them, without the need for any hype, that the book, with its Epoch 2000.0 maps, would still be selling well into the next century. On the back panel of the jacket we printed a charming extract from James A. Michener's novel *Space*, a fictionalized history of the American space programme, published in 1982:

> 'This is one of the loveliest books in the world,' the professor had said, still clinging to the large flat volume. '*Norton's Star Atlas*. Half the great astronomers living in the world today started with this as boys.'

Before I leave the subject of trade book publishing I want to mention a book that I had absolutely nothing to do with, the UK edition of *Chronicle of the 20th Century*, published by Longman in the autumn of 1988. For those unfamiliar with the book, each large-format page is devoted to an individual month, from January 1900 to December 1987, with coverage of key news events presented as newspaper-style stories, as if written at the time, illustrated with contemporary photographs. I am including it because of its association with the power of a simple anecdote to change one's mind in an instant. An advance copy was on display at the summer sales conference that year, having been left after an earlier session in the room my colleagues and I were about to occupy, and I confess that after an initial glance I felt rather snooty about the book. What on earth was Longman doing, getting involved with such lightweight fare? And would people really pay just under £30 for it? Later that day came the story that at a stroke converted me. One of the sales reps related how he had pulled into a petrol station quite close to his home, at the end of a week on the road. Having filled his tank he suddenly realized that his wallet was all but empty. Inspiration came to him: he explained that he lived locally and would be back inside ten minutes to pay. Meanwhile, he had something 'worth £30', more than enough in those days for a tank of petrol, to leave as collateral. His advance copy of *Chronicle of the 20th Century* was accepted. On his return he found all of the staff crowded into their kiosk, immersed in the book. 'We'd like to buy it,' said one of them. '*Now*.'

Published at the end of September 1988, *Chronicle of the 20th Century* sold out its initial printing within a month, selling 160,000 copies by Christmas and reaching the top of the bestseller lists. I bought a copy at the time and have consulted it regularly ever since.

At the beginning of that year, 1988, we learned that Pearson, Longman's owners, had acquired Addison-Wesley, one of the top college science publishers in the USA, with plans to create an integrated Addison-Wesley-Longman. At last, my science colleagues and I thought, a solution was at hand to the problem that had always dogged us: not having a wholly satisfactory arrangement for marketing and selling our books in America. But the integration negotiations, conducted at chief executive level, dragged on. Our authors were pestering us for news; we ourselves wanted news. Suddenly the curtain began to rise. My boss, Peter Warwick, attended an Addison-Wesley sales meeting in the autumn of 1988 and, on his return, asked me to arrange to go to their main college sales meeting, in Tucson, Arizona, the following January.

I described in Chapter 4 some episodes I witnessed at the Addison-Wesley and Benjamin Cummings sales meeting in January 1989. It was for me a wholly absorbing experience, a window on a new world: seeing at first-hand how one of the best American college publishers planned sales and marketing campaigns for its forthcoming textbook list. I drank it all in, seeing so many things that we could benefit from at Longman, and made many useful contacts amongst my new American colleagues.

The habits formed when I had been a field editor at the start of my publishing career came back to me and I felt the need to plan and write a report, setting down my impressions, for colleagues in the UK. A couple of weeks later, in the canteen at Longman, Robert Duncan, now my boss's boss, told me that he had enjoyed reading my report and had sent a copy to Tim Rix, Longman's Chief Executive and, briefly, chairman of Addison-Wesley-Longman. It was confirmation that a piece of writing had worked, successfully communicating my enthusiasm, and that was pleasing. Alas, it was announced only two weeks later that the planned marriage had been called off: Longman and Addison-Wesley were to pursue their separate paths. And two months after that came the news that Tim Rix would be retiring at the end of March in 1990. It had been an exciting time, though rather a brief one, but normality had now returned.

Meanwhile, my parallel interests in both textbooks and trade books continued and I had my sights set on three of the four basic textbook slots in chemistry, with the ground already prepared—discussions with authors were under way—for undergraduate texts in organic, physical, and general chemistry (Longman already published a well-regarded inorganic chemistry textbook, the fourth of the basic slots, by A. G. Sharpe). It was almost time to prepare a paper setting down my plans for chemistry textbooks but I needed first the latest details on the competing texts. The easiest way to gather the data was to examine the books in a well-stocked academic bookshop and the nearest, in Cambridge, was Heffer's (now Blackwell's). I drove to Cambridge on Friday, 19 October 1990, and settled into the chemistry section at Heffer's to take notes. At one point two browsers entered at the other end of the section and one of them, an American, began explaining to his companion how the books were arranged. Presently, just before they left, the American wondered aloud if his latest book was on the shelves. 'Yes, here it is.' This was followed by: 'Let's make it a little more prominent.' The book was rearranged to a face-out position, and the pair departed. Intrigued, I walked over to identify the book and discovered that its author was none other than F. Albert Cotton. He and the British chemist Geoffrey Wilkinson are the authors of an inorganic chemistry textbook, first published at the beginning of the 1960s; with successive revisions *Cotton & Wilkinson* has been for over half a century one of the best-known chemistry textbooks in the world. How pleasing to witness that the great man was not above giving his latest book a helpful promotional push. Good for him!

I returned to the office in the afternoon and was immediately summoned for an audience with my boss who, without preamble, began by telling me that the decision had been taken to 'reconfigure' the division. The heart sinks. Use of that particular euphemism all but invariably signals that the other r-word is about to be deployed. And so it was: I was being made redundant.

This dramatic development needs some explanation. Behind the decision, I was told, was what was considered to be the disappointing performance of the science list over the years. That in itself was fair enough but the underlying cause was for me the real issue and I set down an analysis of it at the time, not to change the decision but to give me the satisfaction of having it placed on the record. The university division had since my arrival been led by a new managing director every two years, and each one had pursued a different publishing strategy, with markedly contrasting priorities, especially on the science side. In a sentence, the continuity needed in academic publishing for steady list-building had been lacking. I should add here that my working relationship with my new boss was not an easy one and I could see then, and can certainly see now, that I would have been perceived as an obstacle to progress in the post-reorganization division.

Various other, subsidiary reasons were given to explain the bad news and one of them brought back memories of the one responsible for the calling of time at W. H. Freeman: late arrival of typescripts from authors, thus upsetting the budget. I decided it would be lovely to have a job where my survival didn't depend on the prompt arrival of type-scripts by delivery dates agreed with authors when books were origin-ally signed up. The perfect, stress-free job, I immediately thought, would be to become a postman, and I mentioned this romantic notion to my friend and colleague at Longman, Andrew MacLennan. He told me to forget it. 'It's like Eton,' he explained. Postmen, he went on, put down the names of their sons shortly after birth in the expectation that they will eventually join the Royal Mail. 'You won't stand a chance,' said Andrew, and I immediately dropped the idea.

But I remained keenly interested in a complete change of career and so asked Personnel (perhaps, alas, they'd already become Human Resources by then) if vocational guidance courses might be available. It was then, purely by chance, because I'd asked the question, that I heard about outplacement agencies. Longman would be prepared to sponsor me on a programme arranged by one such organization, Sanders & Sidney in London. Would I like to talk to them to see if I thought this sort of programme might be my thing?

A breezy Sanders & Sidney executive called Brian conducted my initial interview. 'Now, your departure from Longman,' began his open-ing, no-nonsense question. 'Was it *personality* or *bottom line*?' I liked the directness but confessed that I hadn't really thought of it like that, and then went on to ask the question that genuinely interested me: what was the proportion of people coming to them who completely changed career? 'Ah,' said Brian. 'Not many of our candidates end up as proprietors of flying clubs.' I took this to mean that I was stuck with publishing, but was happy to sign up to the programme.

The first two weeks were diverting and agreeable, with daily two-hour, one-to-one sessions with my counsellor during which we reviewed my life and achievements: for one evening's homework I had to assemble a list of 15 of these, in order of importance. And there was valuable tuition on handling interviews, including responding to the deceptively simple invitation: 'Tell me about yourself', to be answered in precisely 120 seconds. The climax of this first phase was being given a challenging interview and then watching it played back on video whilst my performance was criticized. The result was a Lesson in Life: I had taken the exercise so seriously, prepared for the examination so thoroughly, that I came over as a robot delivering perfect responses. I was advised to lighten my delivery with occasional dashes of quirkiness. For example, as my counsellor put it, if only I had included in my list of hobbies something along the lines: 'My passion is the architecture of working men's clubs in Leeds.' It was a good suggestion, and memorable too.

My campaign to find a job, with the support and help of Sanders & Sidney, got under way and the breakthrough came in April of the following year, 1991. I had contacted Linda Chaput, Neil Patterson's successor as president of W. H. Freeman, which had by then moved from San Francisco to New York. W. H. Freeman and its parent Scientific American had been bought in the 1980s by the German publishing group, Holtzbrinck, and Linda was about to visit its science-book publishing arm, Spektrum, in Heidelberg. Linda had heard that Spektrum wanted to expand its English-language science-book publishing and she promised to talk to them about the possibility of them taking me on. The final outcome was that I was indeed given a job, to be based at the W. H. Freeman office in Oxford, and from there build a list of books of my own choosing.

CHAPTER EIGHT

BILL HAMILTON
AND
JOHN MAYNARD SMITH
WORKING WITH TWO GIANTS OF
EVOLUTIONARY BIOLOGY

IT WAS STRANGE, and strangely wonderful, returning to my old office at 20 Beaumont Street in Oxford, having vacated it some nine years earlier when I left W. H. Freeman. I already knew Graham Voaden, now managing director in charge of Freeman's UK office, and we got on well: he was easy-going and I enjoyed his company. My brief was to build a list from scratch, to be marketed and sold in the UK and Europe by W. H. Freeman Ltd, the Oxford office, and in North America and elsewhere by W. H. Freeman Inc, based in New York. In meetings with the management at Spektrum in Heidelberg at the beginning of the summer of 1991, I had outlined the sort of list I wanted to grow—a mix of popular science titles, mainstream textbooks in chemistry, and a small number of biology texts centred around evolution, a list that would complement the Freeman Inc list in New York—and the plan was accepted.

I wondered if I had managed to find the perfect job in publishing: there were no meetings, my boss was all but 400 miles away in Heidelberg, and I had the freedom to pursue books of my choice. There was something else, too, that I found liberating: I realized for the first time where I wanted my career to go. Both the OUP and Longman had sent me on management courses but none of them had ignited any real interest in management and finance. I had asked Robin Denniston, shortly after he had joined the OUP as head of the academic division, during what turned out to be my final months at the Press: what should be the next stage in my career? But Robin didn't get back to me on that one. At Longman, things were rather clearer, with what seemed to be a general assumption that a commissioning editor's job was but a staging post on the way to senior management. A ray of hope suddenly appeared at the Addison-Wesley and Benjamin Cummings sales meeting I attended in Tucson in 1989, mentioned in the previous chapter. There I was introduced to the title 'Publishing Partner', invented by Addison-Wesley to recognize a select few senior commissioning editors who wished simply to carry on being commissioning editors, with no desire to ascend the management ladder. Their title gave them the same status, and remuneration, as senior managers. I resolved, when the time was right, to look into what prospect there might be of having a similar system introduced at Longman, but events—my redundancy —intervened. It had taken me long enough to recognize what had been staring me in the face for so long but at last it was clear: commissioning was for me the most enjoyable job in publishing and I had no wish to do anything else.

In the autumn of 1990 a monthly column called 'Molecule of the Month' first appeared in the weekly science page of the *Independent* newspaper. Written by John Emsley, science writer in residence at Imperial College London, the idea behind the series was to describe a chemical compound or element that had just been in the news, in a way that would capture the interest of a lay readership. The first one was about beryllium, chosen because of an explosion at a processing plant producing the metal in the USSR, and it was followed the next month by one on bromine, after a tanker accident on a motorway in the UK had resulted in a spillage of hydrogen bromide. The subjects covered in the column ranged widely, from surfactants, met in everyday life as soaps and detergents (the piece on them prompted by the cleaning of stricken birds after Iraq had deliberately released oil into the Persian Gulf at the beginning of 1991), to sucrose, its appearance in the column triggered later that year by a recommendation of the World Health Organization that the amount of sugar in our diet should be cut. They were well-crafted pieces and I could see how people professing no interest in chemistry would in fact find them engaging to read. They gradually changed my long-held view that chemistry, unlike cosmology and evolution, for example, could not be popularized.

I contacted John Emsley and we met for the first time in June 1992. It marked the beginning of a fruitful author–editor association that went on to produce no fewer than five trade books, written for the general reader, over the following ten years or so. The first arose because John wanted to confront, and try to do something about, the undeniable yet perverse fact that while many manufactured chemical compounds are responsible for our modern standard of living, 'chemicals' were, and are, widely regarded with suspicion, often with alarm. I remember witnessing a small but telling example of this soon after my initial meeting with John. A shopper next to me in a supermarket had been examining rival brands of a canned food on the shelves but then returned empty handed to her partner and children. 'None,' she declared. 'They all contain chemicals.' What she presumably meant was not 'chemicals'—everything is made of chemicals—but 'additives'. The negative image attached to the word 'chemical' was not easily going to be shaken off but John was keen to make a start, and the result was his book *The Consumer's Good Chemical Guide: A Jargon-free Guide to the Chemicals of Everyday Life*, published in 1994.

Sales of *The Consumer's Good Chemical Guide* started off encouragingly, helped by positive reviews, and a few months later they received a major boost when the book was shortlisted for the prestigious Rhône-Poulenc Science Book Prize (1995). Established in 1988 by Copus (the Committee on the Public Understanding of Science, set up in London by the Royal Society and the UK government's Office of Science and Technology) and the Science Museum, the Prize has been awarded annually to the book judged to have contributed most to the public understanding of science. A national newspaper reviewed the prospects of the shortlisted titles the day before the result was to be announced and the presence of Emsley's book in the list was gently ridiculed.

The shortlist invariably includes a quirky outsider and, readers were advised, *The Consumer's Good Chemical Guide* fell firmly into this category. What, asked the writer, was the book doing in the shortlist? 'Winning, actually', was how Tom Wilkie, science editor of the *Independent*, put it in his report on the result announced at the previous evening's ceremony. It was the first time that a chemistry book had won. Rhône-Poulenc, sponsors of the Prize, ordered a reprint of 9,000 copies and—their practice each year—presented one to every secondary school in the UK. Paperback rights were sold to Transworld, whose Corgi edition appeared in 1996.

John Emsley's 'Molecule of the Month' appeared on the first Monday of the month, the day when the *Independent* published its weekly science page. With a regular slot on a different Monday during the same period, 1990 to 1997, Bernard Dixon's 'Microbe of the Month' column provided equally pleasing reading. I had known Bernard since the 1970s, when he was editor of *New Scientist*. He approached me at Spektrum about his plan for a book that would portray the activities of a wide range of microbes through a series of 75 vignettes, each focusing on a particular organism. *Power Unseen: How Microbes Rule the World*, published in 1994, was, in short, a portrait gallery illustrating microbial life in its astonishing diversity.

A sceptic unfamiliar with the book might acknowledge that tiny, unseen bacteria, viruses, fungi, and protozoa might indeed pervade every aspect of human society and the natural world. The same sceptic might allow that microbes may have profoundly influenced history, and that they were now helping to shape our future. Even so, wouldn't such a portrait gallery, however worthy, nevertheless be dry and academic? I did actually listen to reactions along these lines when telling friends and colleagues about the forthcoming book. It was satisfying to defend it, though I was never able to match the eventual mouth-watering description of Tom Wilkie, who reviewed the book in the *Independent*. We printed the following extract from his review on the cover of the subsequent paperback edition:

> If Steven Spielberg is looking for a sequel to *Schindler's List*, he could do worse than start with this book.

One key to its success is simply that each individual narrative is so well written. But there is a deeper point: the author has stepped outside the laboratory to engage with the real world. We humans may think of ourselves as the lords of creation, but Dr Dixon shows that the microbes render our tenure insecure. *Power Unseen* is ostensibly a book about microbes. The reason it is so appealing is that, in reality, it is about ourselves.

The books by John Emsley and Bernard Dixon had, by my standards, relatively short gestation periods. At the other end of the spectrum was my association with W. D. Hamilton. In *The Selfish Gene*, Richard Dawkins described and popularized the work of three giants of evolutionary biology, Robert Trivers, John Maynard Smith, and W. D. Hamilton. Reading the accounts of their achievements had so captivated me that I resolved to meet them, and this I managed to do: Bob Trivers at Harvard (my meeting with him is described in Chapter 3), John Maynard Smith at the University of Sussex, and, at the end of 1976, Bill Hamilton at Silwood Park, Ascot, the field station of Imperial College London, where Bill had been a lecturer since 1964.

Over lunch I asked Bill if he had any plans to write a book. Yes, he had, and it flowed from his passionate interest in the huge range of organisms that lived under the bark of rotting wood. It was here, he believed, that the key events in the evolution of insects had taken place, and the title of the book he had in mind was *Dying Wood and the Living World*.

In the event these ideas appeared a couple of years later in the published proceedings of a conference at which Bill presented a paper entitled 'Evolution and Diversity under Bark'. But by this time Richard Dawkins had suggested to me that it would be wonderful to republish all of Bill Hamilton's papers. This was no routine proposal. In terms of the number of publications, Hamilton's output was modest, but each one was hugely important and enormously influential. But because he published in such a variety of places, many of his papers were difficult to get hold of.

Richard had planted a seed and the goal, publishing Hamilton's collected papers, was to remain in my sights for the next 15 years. My next meeting with Bill Hamilton was during the autumn of 1979, by which time I was with W. H. Freeman and Bill was at the University of Michigan at Ann Arbor. I had arranged to visit him there during a trip to America and we discussed the collected papers idea. Bill was typically modest—was it really worth doing?—but said he was happy for me to proceed with a formal publishing proposal. It wasn't difficult to get a range of well-respected academics to support the proposal and W. H. Freeman was happy to commit itself to publishing the collection. In the proposal I noted that each paper would be accompanied by 'short, linking commentaries', provided by Hamilton. He decided he would prefer not to sign a contract at that stage, not wishing to be distracted from his research—there was nothing more to it than that— but we stayed in touch, with my visiting him in Ann Arbor again at the beginning of 1981. I organized a further meeting in 1983, with a separate one arranged with Richard Wrangham, a British primatologist I had got to know when he had been at Cambridge a few years earlier. Richard was now at Ann Arbor, and I was with Longman. Richard gave me the news on my arrival that Bill had 'done the absent-minded professor thing', forgotten about my visit and a few days earlier had left Ann Arbor on a short trip.

Bill left Michigan in 1984, joining the zoology department at Oxford, where I occasionally visited him. Then, towards the end of 1993, Richard Dawkins telephoned to tell me that the collected papers idea, dormant for so long, had suddenly burst into life: Bill was now keen to press ahead. But—the reason for the urgent call—there was a danger that I might lose it. During a recent trip to America, Bill had been discussing his planned book with the American anthropologist and primatologist Sarah Hrdy, editor of a series on Evolution and Human Behaviour published by Aldine, a small academic publisher. Bill, it seemed, was ready to sign a contract. Richard was as aware as I was that the book needed a publisher that could reach a general market as well as an academic one. Both of us believed this because we had read some of the essays Bill had already drafted as his intended introductions to the papers. A book of collected academic papers would have been one thing but the inclusion of Bill's substantial, intensely personal autobiographical pieces, in prose wholly accessible to the lay reader, made it something completely different.

I arranged to see Bill in the zoology department right away and set about explaining why I was convinced that the book had a much wider market than he imagined. Touchingly surprised to hear my upbeat assessment of the book's prospects, he acknowledged that it made sense to go with Spektrum because this would mean marketing and selling throughout the world by W. H. Freeman, a publisher able to reach both academic and trade markets. I promised to organize the publishing formalities with all speed, with the intention of getting a contract to him for signature before he departed in a few weeks' time for a three-month visit to Harvard.

By this time Bill and I had decided to divide the book into two volumes. The first, *Evolution of Social Behaviour*, would contain all of Hamilton's publications prior to 1981, a set especially relevant to social behaviour, kinship theory, sociobiology, and the notion of 'selfish genes'. It would include several of the most read and famous papers of modern biology. A further volume, to be prepared after publication of the first, would be devoted to the second half of Hamilton's life's work, the evolution of sex. And Bill had chosen the title for the book: *Narrow Roads of Gene Land*.

The contract was drawn up and posted to Bill for signature, but the feelings of relief on winning a book I had been pursuing for so long, having almost lost it, were short lived. I rang Bill a few days later to check that all was well only to learn that he had left for Harvard earlier than planned. Might Sarah Hrdy, the academic series editor in America who had previously talked to Bill about placing the book with Aldine, try to change his mind, and succeed? There were no contact details for him and so I would have to wait for three months to find out. Bill was due back in Oxford towards the end of February 1994, and I tried telephoning him then, but without success. Worry that all was not well increased and by the end of the morning I decided to walk round to the zoology department. The lift door opened on the animal behaviour floor and by chance Richard Dawkins happened to be passing. He greeted me and I explained that I was looking for Bill Hamilton.

'Ah,' said Richard. 'Now is not a good time to disturb Bill.' He explained that Bill was required to give one lecture in the year, and today was the day: he was at that moment busy making final preparations. 'Why not go along to it yourself? Lecture theatre B, two o'clock.'

I had assumed without thinking that Bill's lecture would be for staff and postgraduates but realized as soon as I reached the lecture theatre that it was for a large class of undergraduates. I joined the line of students filing in—trying to look inconspicuous—and sat near the back. The large theatre was all but full and Bill began. His reputation as a terrible lecturer was widely known (see Notes and references) and listening to his quiet, monotone voice it was soon easy to understand why. Even so, being aware also of his reputation as a colossus in his field, I became engrossed: here were some of the grand ideas I'd previously read about, coming from the horse's mouth. Many of the undergraduates seemed rather less impressed and when Bill announced after an hour, the time allotted for the lecture, that he wasn't quite finished but would understand if some might wish to leave, a sizeable contingent did just that. For me, the climax of the lecture came towards the end. Bill was discussing sexual selection and how the females of a particular species of bird examine potential mates for visual evidence of the absence of infection by parasites. I can still see projected onto the screen a large image of a bird with a prominent, spotless breast, and can still hear Bill's animated peroration: 'Ladies and gentlemen, I can think of no better way for a male to signal to females his freedom from disease than the display of this immaculate breast.'

After the end of the lecture Bill dealt with questions from a few remaining students and then at last we were alone. The signed contract was ready upstairs in his office, he said immediately. He had posted it back to me before leaving for Harvard but had used the wrong address and it had been sent back to him, remaining in his in-tray until his return three months later. I got back to my office in a state of exhilaration, asking Liz Warner and Jana Bek to join me so that I could tell them the tale. An American, Liz had joined W. H. Freeman the previous summer as managing director of the Oxford office; Jana was its marketing manager. Liz was much taken with the book's title. 'Narrow Roads of Gene Land,' she intoned with a faraway look after I'd finished relating the saga. Then: 'How cool.' I was pleased with that particular reaction. It meant that W. H. Freeman, in Oxford at least, would be wholly behind the book when it was eventually published, two years later, in 1996, twenty years after my first meeting with Bill Hamilton. The reviews, appearing in both scientific journals and broadsheet news-papers, were wonderful. A sentence from the one by Mark Ridley in Nature provides a taste: 'The papers in this book are already classics of scientific research, and the introductions deserve to become classics of scientific autobiography.'

Mark Ridley was right: Bill Hamilton had invented a whole new genre of scientific autobiography. As Alan Grafen later put it to me, Bill had created an amazing vehicle in which he revealed himself, his science, and his times. Alan, now Professor of Theoretical Biology at Oxford, was a close colleague of Hamilton.

I had visited John Maynard Smith, whose work, like that of Hamil-ton, had been described and popularized in The Selfish Gene, not long after it had been published in 1976, and though I enquired whether a book might be in prospect, the time was not ripe. In the event, the ripening process, my chance to work with John Maynard Smith, took rather more than a decade. My author Walter Gratzer had suggested shortly after my arrival at Spektrum that I should approach Maynard Smith about the possibility of his writing a biography of J. B. S. Haldane (Maynard Smith had been a student of the illustrious Haldane). John replied promptly to my letter. The biography didn't interest him but a new book on evolution did: was this something I would like to discuss? The eventual outcome was The Major Transitions in Evolution by John Maynard Smith and Eörs Szathmáry, published in 1995.

Three memories from that time have stayed in my mind, two of them pleasing, the other less so. The first one is of my visits to the Uni-versity of Sussex for periodic meetings with John to discuss progress with the book. These were delightful occasions: always interesting, and always fun. Several members of John's research group would join us when we'd finished talking about the book and we would make our way to the Swan, a pub just off the campus, for a sandwich lunch washed down with a pint of King & Barnes's Sussex bitter.

Regularly visiting authors during the writing stage was a habit I traced back to my time as a field editor. Authors liked it, having their editor come to see them to talk about progress, and I found it useful, hearing at first-hand how things were going. It was a lubricant to progress, and a way of keeping in touch with what was going on in a particular field, and in the academic world in general. In this particular case, there were no special problems that needed to be resolved. We talked about what he and his co-author (in Hungary) were currently writing, what was to be tackled next, and how the overall timetable was coming along.

A second recollection was printed in the eventual book and it concerns an exchange between the authors and one of the academic reviewers I had lined up to comment on the draft chapters. The specific matter is arcane but that doesn't hide its amusing side. In their Preface, the authors thanked the reviewers in the customary way, especially for suggesting ways in which things could be explained more clearly. John and Eörs continued:

> We have not always taken their advice: it is impossible to resist mentioning one example when we did not. We drew a parallel between Eigen's notion of an error threshold and a phase transition. One of the three [reviewers] wrote, 'This does NOT remind me of a phase transition: it cannot be too widely known that nothing reminds me of a phase transition.' Believing, as we do, that the only thing required to make this particular commentator the complete evolutionary biologist is a love of phase transitions, we have left the remark as it stood.

The final reminiscence is of things going wrong. A routine had been established whereby W. H. Freeman's New York office completed a standard sales projection form for my books. The form for *The Major Transitions in Evolution* was sent to me by Bob Biewen, the president of Freeman, and it forecast a life-time sale of 700 copies. Appalled, I wrote to Bob and said that my boss and I believed that their projection very seriously underestimated the book's potential, going on to advise him that I was withdrawing the book and that I would attempt to place it with another American publisher. Bob subsequently telephoned to confirm that they would be pulling the book from their forthcoming sales meeting. Meanwhile I had contacted W. W. Norton in New York— I had collaborated with them on *The Blind Watchmaker*—and, following an immediate expression of interest, sent them a set of proofs. Less than an hour after despatching them, Bob Biewen called me from New York. He had at last talked to his biology editor, who confirmed that Maynard Smith was a big name. Please would we change our mind and let them have the book, given that they would raise their first order to 2,500 copies? After pondering over the weekend I agreed to their request but subsequently lived to regret it: Freeman in New York forgot to plug the book into their systems and their initial sales were disheartening.

It is time to draw the story to a close with a postscript. At the time of writing, *The Major Transitions in Evolution* is still in print and still selling, more than 15 years after it was published.

In the previous chapter I mentioned Marian Dawkins's *Unravelling Animal Behaviour*, a textbook for student courses published by Longman in 1986. The choice of subject for her next book, animal consciousness, was especially appealing because there would be a general market as well as a student one. The possibility of conscious experiences in non-human animals was one of the great remaining biological mysteries. What goes on inside the minds of other animals? Do they have thoughts and feelings like our own? In *Through Our Eyes Only? The Search for Animal Consciousness*, published in 1993, Marian set out to describe the results of recent research in animal behaviour and argued that the idea of consciousness in other species had by then progressed from a vague possibility to a plausible, scientifically respectable view.

My American colleagues at Freeman in New York liked the book after reading it in typescript and proposed that if we agreed to change its title to one they had in mind they would be able to increase their order by several thousand copies. Marian and I chuckled over their whimsical choice, *Dolphins at Dreamtime*, but decided to stick with the original title, not least because dolphins didn't feature anywhere in the book.

Sometimes, though, suggested title changes from America could result in a real improvement. The size of the serious end of the amateur astronomy market had been brought home to me whilst working on *Norton's Star Atlas* at Longman, and at Spektrum I signed up James Muirden, one of the contributors to *Norton's* 2000.0, to edit a handbook for the advanced amateur astronomer. The working title was the straightforward *Practical Amateur Astronomy* and it didn't occur to me that something less prosaic might bring benefits. Colleagues at Freeman in New York felt otherwise and came up with *Sky Watcher's Handbook*. James Muirden and I were more than happy to make the change and the book was published in 1993.

Titles have rarely been a problem when dealing with American co-publishers, but where jackets and covers are concerned, the dramatic difference between British and American tastes occasionally bemused me. One example illustrates not only this difference but also the problems one could encounter when publishing a book that didn't fit neatly into one category. In 1995 I published Richard Turton's *The Quantum Dot: A Journey into the Future of Microelectronics*. Like Marian Dawkins's *Through Our Eyes Only?*, Richard's book was so well written that it succeeded in appealing to both the trade and student markets. In the UK, *The Quantum Dot* was taken by The Softback Preview book club (a modest 1,500 copies, but very welcome) and at the same time was recommended for undergraduate courses in physics and electrical engineering. But this so-called crossover versatility seemed to perplex my American marketing colleagues at W. H. Freeman and they decided that the book was not for them. From the New York-based headquarters of Wiley, a mighty presence in science publishing, came the confession that *The Quantum Dot* had had them scratching their heads. 'The trade department think it's a college title, while the college department think it's a trade title', I was told. My patience was rewarded when I finally tried OUP USA and got the thumbs-up from their physics editor, Jeff Robbins. And Jeff liked the cover we were proposing to use, this giving me a special pleasure on account of the story behind its creation.

My books, jackets, and covers were being designed by the exceptionally talented freelance designer, Pete Russell. Intrigued by the contents of the Turton book, he devoured large portions of the typescript, going on to quiz the author, asking what quantum dots actually looked like. The result was the arresting cover illustration—depicting the atomic world of the quantum dot, the likely basis of electronic devices in the future, each one of them several hundred times smaller than the size of an average dust particle—that Jeff Robbins at the OUP in New York liked so much. But Jeff didn't have the final say. Presently I received a note from his marketing colleague, Amy. 'I am countermanding Editorial's approval of the UK cover for our edition', the note began. I telephoned to assure Amy there would be no problem in supplying them with books with their own jacket; and then, finally, wholly intrigued, I asked her what was wrong with our cover. Her tone communicated astonishment that I was asking a question with such an obvious answer. 'Oh, it's too British', was the matter-of-fact reply (see Plate 6).

The jacket of the American edition of *The Quantum Dot* was, to my eye, exceedingly boring but the OUP had soon sold the 4,000 hardback copies we had supplied, going on to publish their own paperback edition. *Chacun à son goût.*

In April 1994 I attended the annual dinner of the National Secular Society, having been asked to deliver a short speech and then propose a toast to their guest of honour, Peter Atkins. The invitation had come out of the blue and I wrote to the Society asking what sort of speech they wanted. 'We like the toast to the Guest of Honour to be a sort of general statement of what a nice chap he is and any amusing stories about him. We like anything which has a dig at religion . . .'.

I told my audience about Peter's book *The Creation*, published in 1981 when I was with W. H. Freeman, which had set out to explain how the universe came into being without the need for a creator. (Peter subsequently updated the book and his *Creation Revisited* was the first title I published with Spektrum, in 1992.) The book had recently been in the news, I went on, attacked by Bryan Appleyard, the author of *Understanding the Present: Science and the Soul of Modern Man* (1992), in the *Independent* a few weeks earlier. The piece had attacked science— science does not leave room for the soul—and referred in uncomplimentary terms to Peter Atkins and 'his dreadful book *The Creation*'. There had been a swift retaliation from Richard Dawkins and I quoted from his letter to the *Independent*: 'Peter Atkins's *The Creation* is not a "dreadful" book, it is perhaps the most beautifully written work of prose poetry in all scientific literature.'

I went on to talk briefly about the soul, guessing that the subject would interest the National Secular Society diners, beginning by recalling a conversation over lunch with Peter Atkins the previous summer. Peter had told me about Channel 4's plan to televise two debates on the existence or otherwise of God, each to begin with addresses from two opposing speakers. One of the debates was to pit Peter against Richard Swinburne, then Professor of the Philosophy of the Christian Religion at Oxford. I explained that Peter had gone on to tell me that Swinburne claimed to know the shape of the soul. 'You mean metaphorically?' 'No,' replied Peter. 'Literally.'

My story moved to Cambridge, where I happened to be the week after lunch with Peter. One of my authors I had called on was John Polkinghorne. At the end of the first book we had worked on together, *The Particle Play: An Account of the Ultimate Constituents of Matter*, published by W. H. Freeman in 1979, John had recorded that it had been written while he was Professor of Mathematical Physics at the University of Cambridge, but publication would be when he was an ordinand studying for the Anglican priesthood. John and I went on to collaborate on two further books, published at Longman: *The Quantum World* (1984) and *Rochester Roundabout: The Story of High Energy Physics* (1989). During the meeting with John in 1993 I asked about Richard Swinburne and the shape of the soul. Yes, said John. He's calculated that it is spherical. That piece of information vouchsafed, it was time to propose the toast to the Society's guest of honour.

Richard Dawkins had given the televised Royal Institution Christmas Lectures in 1991, with the title 'Growing Up in the Universe', and planned to develop them into a book. His literary agent was now the formidable John Brockman, whom I met in London in early 1992. The meeting had been arranged after Richard had told John that he was keen to work with me on the new book. John had become well known for the sizes of the royalty advances he was able to negotiate for his authors and it soon became clear that the next Dawkins book was out of my league. It was snapped up by Viking, the hardback imprint of Penguin, and shortly afterwards I was contacted by Ravi Mirchandani, then science editor at Penguin, and it was agreed that I would edit the book on a freelance basis (this having been cleared with my boss in Heidelberg).

Initially intended as a book based on the Lectures, it soon started to change. Part of a letter I sent to Richard (in April 1993) gives a taste of this process of evolution. I had begun with some negative criticism of what, in the original version, was Chapter 2 and went on to comment generally on the chapter that followed, beginning with 'An excellent opening . . .' This was followed by:

> I read on, a little apprehensive after my criticisms of the previous chapter, and gradually (though not too gradually) the dawning realization that this is wonderful. This is more like the author of *The Selfish Gene* and *The Blind Watchmaker*. This is vintage Dawkins!
>
> Why? The narrative unfolds beautifully: a strong thread as you explore one aspect and then move smoothly to the next. This gives it a unity and elegance. The ideas and the way they're developed and explained are deeply interesting: I was genuinely gripped. And there are the animal stories —the pigeons, the moths and the predator birds, and the angler fish— which breathe life into it all.
>
> You must wonder what to make of me, swinging from one extreme to the other with these two chapters. I suggest that this chapter is an example of your natural territory. That's not to say that Chapter 2 can't be got right: of course it can. But it raises the general question of level. You yourself said when you gave me Chapter 3 that the level was going up, and of course you're right. But this is your natural level and staying there is going to result in a better book. It means thinking about how to pitch the two opening chapters, but my advice is to leave them aside for the moment, pressing on and coming back to them later.

Back to Chapter 3, you will see that I have marked just one (short) passage, halfway down page 14 to near the top of page 18, where the pace of the explanation needs slowing down.

Eventually Richard reconstructed the book, giving it a new title, *Climbing Mount Improbable*, and it was published (by Viking in the UK and by Norton in the USA) in 1996.

In November 1996, I received a letter from Heidelberg telling me that Spektrum's UK operation was alas going to be closed down. Of my various notices of redundancy over the years, this was by far the most sensitively delivered. The reason for the decision was that student numbers in Germany in the sciences, particularly in biology and chemistry, had unaccountably plummeted. This was having an immediate effect on Spektrum's domestic textbook sales and Holtzbrinck, the parent company, had instructed them, among other things, to sell the list of its Oxford office, and then close it down. My boss stressed that no specific timetable was being imposed. Moreover, they would need my help in finding a buyer for my list.

I contacted the obvious candidates who might be interested and one of them, the OUP, wanted it more than the others. The OUP offered me a one-year freelance contract and I moved into my new office there in June 1997, just over 18 years after I had left the Press in 1979.

CHAPTER NINE

'THE BEST TEXTBOOK OF ORGANIC CHEMISTRY I EVER HOLD IN MY HANDS'

OVER THE YEARS I listened to many a valedictory speech at leaving parties at the OUP and without exception the speakers all singled out one thing they would miss: being surrounded by such a stimulating group of colleagues. They were all spot on in this regard, and I felt the real buzz of being back at the Press from the first day of my return.

The purpose of my one-year freelance contract was to enable me to 'deliver important copyrights', as the management put it. This was an allusion to books signed up at Spektrum and still being written, some of them close to completion. In addition, I was given clearance to commission new books: the same freedom I had enjoyed at Spektrum to pursue books of my choice. There were no hints about the possibility of survival after my year was up but, if there were to be any chance of eventual salvation, my new-commissioning record would presumably be one way to influence and help bring about that happy outcome. So although my position was precarious, it was also a splendid, exhilarating time.

The books selected for mention in this chapter were contracted at Spektrum and eventually published by the OUP. By far the most important, measured in terms of potential sales and income generation, was a mainstream textbook of organic chemistry. For me, it had long possessed all the hallmarks of a classic in the making. It had started life as an idea 15 or so years earlier, when I was with W. H. Freeman, during a lunch with John Mann, an organic chemist I had known from the 1970s, when he wrote a volume for the OUP's Oxford Chemistry Series.

I had better spell out briefly the reasons why there was such a (tantalizingly large, but by no means straightforward) market for a new organic chemistry textbook in the 1980s. British, European, and American university bookshops typically stocked piles of American organic chemistry texts, substantial tomes of a thousand pages or more, each written and finely tuned for the big sophomore (second-year undergraduate) organic chemistry courses in the US. Sophomores would all have completed in their previous year a freshman 'general chemistry' course, and after obtaining their credit in organic chemistry, most would take no further chemistry courses. The European system is totally different: there is no freshman year of general chemistry (though some of this material is taught in school). The result of this difference is that all of the American organic chemistry texts are broad but shallow in their coverage, and that, in the UK and Europe, results in these books being too broad for our first-year courses, and too shallow for our second-year ones. There was a widespread feeling, too, that the American texts tended to be 'safe', meaning traditional and dull. But British and European universities chose an American organic text because they were the only ones available. All of these universities, John Mann assured me, would like a text written specifically for their courses rather than for the very different one given to American sophomores.

I apologize if the above detail sounds a bit tedious to the general reader, but such considerations were central to my publishing role, and I certainly found them totally absorbing.

Who might take on the daunting task of writing a British textbook of organic chemistry? John Mann had no doubt who the ideal author was: Stuart Warren at Cambridge.

I arranged to see Stuart not long after and we had lunch at the Coach and Horses pub in Trumpington, on the outskirts of Cambridge. Yes, said Stuart, there is a real need for such a book. Moreover, he would love to take it on, if only he had the time. We left it that I would call on him during future visits to Cambridge in the hope that he might eventually feel able to say yes. I did just that over the years, popping in for a cup of coffee and chat, during successive spells with W. H. Freeman, Longman, and Spektrum. At some point—a glimmer of hope that things might eventually start to move—Stuart proposed taking on a co-author, Nick Greeves, to be ready when the time was ripe. Nick had been a graduate student of Stuart's at Cambridge and was now a lecturer at Liverpool University. I made a point of calling on Nick whenever I was in Liverpool.

Then, in March 1994, the long-dormant project suddenly burst into life. I had arranged to see Stuart for what I expected would be a routine meeting, but realized on entering his office that things were different. More animated than usual, he described the moment when he'd realized that what was needed was the addition of two more co-authors, Jonathan Clayden and Peter Wothers. Jonathan was another former graduate student of Stuart's who had just been appointed to a lectureship at Manchester University. Peter had only just completed his doctorate at Cambridge and would bring to the team an immediate knowledge of what the current generation of students was like, and what aspects of the undergraduate course they found difficult.

Stuart subsequently drafted a synopsis for the book, together with an introductory note explaining why this new textbook would be different, and in August of that year, 1994, two day-long meetings were held, the first in Cambridge and the second in Manchester, when the draft synopsis was debated, line by line. It was then, sitting in on these meetings, that I became convinced the eventual book would be a winner.

A final proposal went out—for comments and endorsement—to university chemistry departments in the UK, Europe, and the US, and the book was signed up in 1995. The writing took four years, during which the authors and I met three or four times a year to review progress and discuss the latest batch of completed chapters. In the summer of 1997, shortly after the OUP's purchase of the Spektrum/Oxford list, I joined authors, wives, partners, and children for a week in a large rented holiday home in Herefordshire. This had been Stuart's idea, and a full working session on the book each day produced the substantial progress needed as we approached the final phase of writing.

Each evening's dinner during that memorable week was planned, prepared, and cooked by a different, unrelated pair. I had taken my camera and, during breaks in the garden throughout the day, shot informal portraits of individuals and of groups (see Plate 7). One of these, featuring the four authors, was printed on the back cover of the eventual book.

Organic Chemistry by Clayden, Greeves, Warren, and Wothers, referred to by convention simply as *Clayden*, the first name in alphabetical order, was published in July 2000. With 53 chapters and over 1,500 pages long, and with full-colour presentation throughout, it was a bargain at £29.99 (a price of more than £30 would have made the competition —the various American texts which had been available for so long— appear temptingly inexpensive in comparison).

How was it received? Three contemporary snapshots from the time —each from a different angle of teacher, reviewer, and student—provide a taste.

Promotion copies of textbooks intended for course adoption are despatched to academics teaching the course, along with a request for a verdict on the book's qualities and suitability for course recommendation. My favourite of many such comments on *Clayden* communicates glorious, over-the-top enthusiasm. It came from a lecturer at the University of Mainz, Germany.

> Yes, I received it and am really enthusiastic about it! This is by far the best textbook of Organic Chemistry I ever hold in my hands! I recommended it immediately to all of my students and they ordered the book instantaneously, because they also realized this book to be an extremely efficient means of learning organic chemistry. It is really the book I was always hoping for! Chemical concepts, curly arrows, mechanisms, all the tools necessary to organize the enormous amount of material instead of just learning the facts. This was always a maxim in my courses and always the way I taught chemistry. So this book really is the perfect match. At the moment we are about to move from Mainz to Darmstadt and one of the first things I have to do there is to teach organic chemistry at the

undergraduate level. Of course I will use this book and I will strongly recommend this book to replace 'all the other stuff'.

The second snapshot is from a review in the *Times Higher Education Supplement* (now the *Times Higher Education*) by Andrew Boa, lecturer in organic chemistry at Hull University. The caption to a picture accompanying the review, a stylized bra, read: 'Cups overfloweth: the range of materials covered is vast.' The other thing that made me smile was the following extract from the review:

> A colleague recently told of an encounter he had with a sales representative from a publisher of one of the standard US texts. My colleague brandished *Clayden* and proclaimed: 'Have you seen this? It's brilliant.' The rep sighed, shrugged her shoulders and replied: 'I've heard a lot of people say that.' But I do not expect my colleague brandished the book for too long. Tipping the scales at just over 3kg . . .

The final snapshot is a letter from an undergraduate to the authors, sent to them via the OUP. It is the only unsolicited letter from a student commenting on a textbook that I saw during my entire career in publishing.

> I am a student at the University of Liverpool studying Pharmacology. I am writing in order to compliment you on your book. I have had difficulties with my chemistry modules in the past due to lack of basics. I failed my first year, redid it and I still had re-sits this September. I couldn't find a textbook that I could learn from . . . it was just facts and more facts and I need to be able to understand the work rather than memorize it. Over the summer I bought your book and I found that it fulfilled my needs completely. It is perfectly written for someone who knows nothing about organic chemistry and dissects each part into easily accessible chapters. I promised myself that I would write to you commenting on the book if I passed, which I did, and I am sure I will need to refer to it again this year so I don't think I'll be having problems with chemistry again thanks to you.

It was a special pleasure circulating copies of her letter to my sales and marketing colleagues, especially to those who had not been entirely persuaded by repeated assurances that *Clayden* was not an over-sophisticated book, written only for gifted students by clever but out-of-touch authors.

The first edition of *Clayden* remained in print for over 11 years, selling 120,000 copies. A second edition was published in the spring of 2012. My personal guess is that with regular new editions it will still be selling in 25 years' time.

The management of a project as huge and complicated as *Clayden* —turning the authors' final version into an edited and printed book— benefited greatly from the considerable resources and expertise available at the OUP. It would have presented a daunting challenge for my assistant and me, sole occupants of the office at Spektrum. Exactly the same points apply to Colin Tudge's amazing *The Variety of Life*, published in 2000, so beautifully written, the broad sweep of science so meticulously researched, that it appealed to an audience comprising general readers, students, and professional scientists. And of all the books I ever worked on, it has my favourite subtitle: *A Survey and a Celebration of all the Creatures that Have Ever Lived*. (My marketing colleagues in New York emailed me when they were finalizing their publicity plans for the book. The subtitle provided quite a selling point, they began. 'Is it actually true?' 'Yes, it is. Go ahead and exploit it!')

Reviewers of *The Variety of Life* ranged from Harvard's Edward O. Wilson ('. . . an easily accessible reference work to the entire spread of biodiversity') and John Maynard Smith ('It contains all the knowledge I hoped to acquire from a degree in Zoology—and didn't') to Ian McEwan, writing in the *Observer*'s *Summer Reading Special* ('I'm going to Spain for my holidays . . . I'll take Lorca . . . Michael Frayn . . . plus Colin Tudge's *The Variety of Life* which is a celebration of all the creatures that have ever lived').

Another complicated book, though it began life as a modest, straightforward one, was proposed by John Emsley as a popular reference work describing each of the hundred or so chemical elements that are the basis of everything in the universe. The entry for each individual element was to comprise five sections: historical (its discovery), economic (industrial production, known reserves, main producing countries), environmental (quantities in the earth, oceans, and atmosphere), chemical details (basic physical data), and the human dimension (the amount of the element in the human body, where it is located, and its function or effects). The last aspect was seen as having the greatest general appeal and so the working title chosen for the book was *The Human Element*. It was signed up in the autumn of 1992 as a 250-page paperback, to be published two years later.

Progress with the writing was uneven, with growth spurts punctuated by periods of dormancy, and it came to the OUP still unfinished, and still growing in extent. An issue for me was the title and I was keen to make use of the Oxford name, with a change from *The Human Element* to *The Oxford Book of the Elements*. What I didn't appreciate was that the perception of the Oxford name (as in *The Oxford Book of . . .* of which there were many examples) had changed during my 18-year absence from the Press: the formula now signalled an anthology rather than an original work, and I was talked out of the idea. One colleague came up with the final title, *Nature's Building Blocks*, and another with the subtitle, *An A–Z Guide to the Elements* (shooting down my final attempt to use *The Oxford Book of the Elements* as the subtitle).

The design department produced a stylish image for the jacket and I was immediately struck by what seemed to be a similarity in the type font used for the word *Nature's* in the title and the one adopted for as long as I could remember by the journal *Nature* on its front cover (although I could see no particular advantage to the book from any such association). But, colleagues assured me, there were no grounds for my disquiet and the jacket was approved. Shortly after the book was published (2001), Richard Charkin, then head of Macmillan, publishers of *Nature*, wrote to the OUP demanding a royalty for the use of *Nature's* brand image. A short correspondence at a senior level ensued and the claim was withdrawn. I called on Richard at Macmillan a couple of months later—we used to meet up occasionally for a chat—and we resolved the matter cheerfully over a bottle of Chablis from his office fridge.

Another notable individual who came into the *Nature's Building Blocks* story was Oliver Sacks, author of *The Man who Mistook his Wife for a Hat*. I met Oliver over dinner at Lincoln College in Oxford, in July 2000, when we were both guests of Peter Atkins, and I learned on that occasion that he had a passionate interest in books on the chemical elements. When John Emsley had finished writing his final entry, on zirconium, I sent Oliver a copy of the complete typescript, hoping that he might feel able to provide a comment we could use. The result was a wonderful endorsement, which we printed on the jacket:

> . . . a marvel—encyclopaedic in scope, but so full of enthusiasm, so engag-
> ingly written, that one can open it at any point and read for sheer delight.
> While everything which one might expect and hope for is here—the
> history and mineralogy and manufacture of each element, its economic
> roles and uses, its place in the environment, our foods, our bodies—there
> is, in addition, an 'element of surprise' for each element—an idiosyncratic
> piece of information, sometimes comic, sometimes terrifying, but always
> startling and novel. I have read and possess many books on the elements,
> but it is Dr Emsley's new book which will now sit next to me on my desk.

Nine years elapsed between the book's being signed up as a 250-page paperback and its eventual publication, initially as a hardback of 550 pages. The hardback sold around 6,500 copies and at the time of writing the subsequent paperback, published in 2003, was still selling, with sales approaching 35,000 copies.

In March 1999, *Nature* invited me to write a piece for a forthcoming spring books issue. Two special issues, published in the spring and autumn each year, carried more book reviews than usual, and occasionally an essay by an outsider, on some aspect of the book world, introduced the reviews section. A number of *Nature*'s readers were interested in the possibility of writing a popular science book themselves, I was told. How should they go about it? How should they choose a publisher? And what does a publisher's editor actually do?

At the time, rather than tackle these and related questions in an abstract, generalized way, I thought it would be a good idea to base the article on my experience of working on a specific project. This would provide a more focused context (and one, I felt, of genuine potential interest), and give the reader a more immediate sense of how parti-cular books emerge from the commissioning and editorial process, while any practical advice emerging from it would very probably carry more weight than editorial hints presented in isolation. I had one particular book in mind for this role—one that had been published that very month—and, although it was by no means a typical project, I thought the unusual story behind it could have an additional appeal for many readers.

Unfortunately, when submitted, my article was promptly rejected—on the grounds that it could be seen as professionally self-serving, not least by providing free advertising for the book. I was disappointed, of course, but, looking back, I can see that they had a point! All the same, the piece seems as relevant now as it did then, and as central to the concerns of the present volume, so I am including it below. (Let me at once reassure readers that, as the book that generated it is currently out of print, no clash of interests is now involved!)

The book concerned is *The Art of Genes: How Organisms Make Themselves* by Enrico Coen, published in 1999. As the jacket blurb began: 'How is a tiny fertilised egg able to turn itself into a human being? How can an acorn transform itself into an oak tree? Over the past twenty years there has been a revolution in biology. For the first time we have begun to understand how organisms make themselves: the mechanisms by which a fertilised egg develops into an adult can now be grasped in a way that was unimaginable a few decades ago. *The Art of Genes* is the first account of these new and exciting findings, and of their broader significance for how we view ourselves.'

Born in Liverpool in 1957, Rico studied genetics at Cambridge and in 1984 joined the genetics department at the John Innes Centre, Norwich, where he is today, working on the genetic control of flower development in *Antirrhinum* (snapdragon). In 1997, he became an Honorary Professor in Biology at the University of East Anglia, and in 1998 a Fellow of the Royal Society. Rico's book, his first, had during its gestation turned into a triumph, the significance of the feat captured in a few sentences written by Professor Michael Akam, Director of the University Museum of Zoology, Cambridge. We reproduced his verdict on the jacket:

> This is an impressive book. Written by a leading practitioner in the field, it combines profound insight into developmental processes with a lively, non-technical style. It presents an intensely personal view of the subject . . . This

is a truly remarkable achievement in a book that can be read and enjoyed by almost anyone.

Here now is the unpublished piece—recounting part of the story behind the book—which I wrote for *Nature* in 1999. I adopted the convention of using the impersonal masculine pronoun—regarding this, then as now, as no more than a convention.

A scientist decides to write a book describing his field for a general readership. It is his first book: how should he choose a publisher? The one offering the biggest royalty advance? Well, that's a consideration, though not all first-book authors are offered advances. A publisher with a successful track record in marketing and selling popular science books? Yes, that would certainly be a plus point. Another thing to think about is that you will be working with an editor. What does an editor do? What might an author expect from their editor? All editors, all authors, and all books are different. Here is one story.

I received a one-page outline for a popular book on development from Enrico Coen (Rico, when I got to know him) in February 1992. The working title was *Form, Flowers and Flies*, accurate if not exactly arresting. It was changed later that year, when Rico sent me a first draft of the opening chapter, to *The Destiny of Genes* and this was finally changed, some four years later, to *The Art of Genes*. Titles are important and a point for those wishing to write a successful popular science book is that you should make sure you end up with a good one. I visited Rico for the first time towards the end of 1992: I like to meet and get to know my authors, and, to be frank, I was concerned because another publisher had also been approached (having read the first draft of the first chapter I already had the feeling that this could be an important and a successful book, a view that was greatly strengthened when reviewers later reported on two draft chapters).

Over the following five years Rico sent me chapters to read and comment on: slow progress to start with and then an acceleration in 1996 (the first draft of the book was completed at the end of that year) when he had some study leave. We met again in February 1996 and went over eight chapters I had annotated. They are stylish, I tell him, and in a class of their own, but the technical descriptions and explanations are too dense for a general readership. Rico confesses that his natural style is very economical, but he is determined to try harder to be more accessible. One of the central ideas in the book is the drawing of a parallel between the way genes respond to the developing pattern of an organism and the way an artist responds to a painting being created on canvas. Using this analogy, Coen describes how an organism develops through an interactive dialogue in which there is no clear separation between plan and execution. Towards the end of 1996 a letter to me from Rico ends: 'The book has turned out to be a microcosm of its own argument. As you will see [from an enclosed chapter], I have tried to argue that, in many cases, planning something is not done before it is executed—the two go together. It is only by writing the book that I have realised what I wanted to say!'

When I am reading and commenting on draft chapters I am representing general readers: like them I have no specialist knowledge, but am willing to make the effort to understand. When it is a struggle to follow an explanation I point out where it is happening, asking for clarification, a slower pace, a consolidation passage, and so on. All of my comments are to do with content, not with the writing itself. So yes, budding popular science authors do need to be able to write well: do not expect an editor to transform a worthy but dull account into lively and bubbly prose.

Revised chapters start to arrive at the beginning of 1997. In February, Rico writes: 'I am finding it hard to judge whether the readers are still with me or whether they have thrown it down in frustration. My feeling is that if they can get over chapters 3–5, they will begin to see development in a new and clearer light. Is this the way it reads or am I deluding myself?' At about the same time, in a letter to Rico accompanying my comments on a couple of chapters, I write: 'My fascination for this field has grown to such an extent that when I next come over to see you I'd quite like to look around your lab and see some of the physical realities that lay behind these brilliant, if challenging, ideas.'

A revised and polished version of the whole book was completed in March 1998 and this was sent out to reviewers for final technical suggestions, and for views on the book's prospects. It was published in March 1999. During that last year I tell everyone in the publishing house who will be involved with the book why it is so special. This is another part of an editor's job: to be a book's champion inside the house. Much of an editor's enthusiasm for a popular science book he's been working on—and you need enthusiasm to be an effective champion—comes from meeting for the first time exciting new ideas. There was a nice signal, when the reviews of the completed typescript came in, that enthusiasm for the Coen book might not be restricted to newcomers. One reviewer began his report: 'Since [the book] is an introduction to a field in which I have worked for over ten years, I expected to find it rather boring to read. Instead, I found it absolutely fascinating.'

I see from the file that over the years I read and provided comments on 32 versions of chapters, as first drafts, revised ones, and in some cases re-revised ones (the number of chapters in the eventual book reached 18). Seven years elapsed between my receiving the first outline and the book's publication. Is this a typical timetable? No. Was it worth all the trouble? Time will tell, but I will end with the prediction that the book will still be selling seven years from now.

I read the piece again when writing this chapter and, curious about my prediction in the final sentence, decided to check it. The paperback edition finally went out of print in 2008, nine years after publication.

My collaboration with Bill Hamilton continued and, with the publication in 1996 of Volume 1 of his *Narrow Roads of Gene Land, Evolution of Social Behaviour* (described in the previous chapter), work continued on his autobiographical introductions for Volume 2, *Evolution of Sex*. By this stage we had decided to restrict the second volume to Bill's publications up to 1990 (with one from 1991): a third and final volume would follow in due course.

I continued to have regular meetings with Bill and it was during this period, the mid- to late-1990s, that I met and got to know Luisa Bozzi, an Italian science journalist who had become Bill's devoted partner. On one occasion they were attending one of the Darwin Seminars, public lectures on a wide variety of aspects of evolution, organized at the London School of Economics from 1995 to 1998 by Helena Cronin. I have a clear memory of this particular encounter because of a tiny but immensely touching episode. Spotting Bill and Luisa during the mid-session tea break, I joined them. I knew that Luisa's current trip to the UK was a short one and, in small-talk mode, asked her when she would be leaving Bill, expecting a response along the lines: 'I fly from Heathrow at the end of the week.' Bill interpreted my innocent question literally and before Luisa could answer he straightened himself up and, slowly and gravely, said: 'I hope that Luisa will *never* leave me.' Luisa, like me taken by surprise, smiled, let out a heartfelt 'Ah', and gave Bill a hug.

At the end of December 1999 Bill and two companions travelled to the Congo to investigate an unfashionable theory that the HIV virus (the cause of AIDS in humans) originated during lax trials of an oral polio vaccine in Africa in the 1950s. Whilst there it was thought that he had contracted malaria and he was flown back to the UK. Later, in March 2000, he died of complications, aged 63.

I attended a secular memorial service, held in the chapel of New College, Oxford, where Bill had been a fellow, on 1 July 2000. Richard Dawkins delivered a eulogy, and the choir sang 'Happy is the man that findeth wisdom' by J. Frederick Bridge, an anthem composed especially for the funeral in Westminster Abbey of Charles Darwin.

Volume 2 of *Narrow Roads of Gene Land, Evolution of Sex*, was published posthumously in 2001. Some of the autobiographical introductions were much revised by Bill during 1999, and at the time of his death he had left the book in an almost-finished state. It was edited by Sarah Bunney, who had been responsible for editing Volume 1, and Mark Ridley assisted in the resolution of remaining textual and reference queries. We included Richard Dawkins's eulogy as a Foreword, following this with the score for Bridge's anthem. And on the cover we printed an extract from Olivia Judson's obituary in the *Economist*: 'He blew up established notions, and erected in their stead an edifice of ideas stranger, more original and more profound than that of any other biologist since Darwin.' Volume 3 of *Narrow Roads of Gene Land*, edited by Mark Ridley, was published in 2005. It includes the papers from Bill's final years, each accompanied by a personal introduction by the paper's co-author. The subtitle of this concluding volume, suggested by Olivia Judson, is *Last Words*.

In 2002, I commissioned Ullica Segerstråle's biography of Hamilton. Ullica had earlier written her account of the sociobiology controversy, *Defenders of the Truth*, published in 2000 and mentioned in Chapter 3. Her biography, *Nature's Oracle: The Life and Work of W. D. Hamilton*, was published in 2013, nine years after I had retired from the OUP. Several endorsements are printed on the jacket, including this one, from Richard Dawkins:

William Hamilton's name stands above all others in evolutionary biology since the Modern Synthesis of the 1930s and '40s. As John Maynard Smith, with whom he had a troubled relationship, said, 'He's the only bloody genius we've got.' As geniuses often are, he was a complex character and an exceptional challenge for any biographer. Ullica Segerstråle is ideally qualified to rise to that challenge. She achieves a genuinely affectionate yet warts-and-all portrait of her subject, combined with a good understanding of the deep subtleties of his thinking. Those who loved him, as I did, and those who wish to know more of the astonishing originality and versatility of his contributions to science, will treasure this book.

I was keen at the start of my one-year freelance contract period with the OUP to get myself more widely known in the publishing world, in case my tenure at the Press came to an abrupt end. During my freelance days I acted as a science consultant for Felicity Bryan, the Oxford-based literary agent, and did editorial work for various publishers with popular science lists. Felicity put me in touch with Stefan McGrath, science editor at Penguin, and this led to my working with Richard Dawkins again, at the beginning of 1998, reading and commenting on his draft typescript for *Unweaving the Rainbow*.

My OUP contract was due to come to an end in the summer of 1998 but I learned as the deadline approached that it was being extended by a further six months. Alan Singleton joined the Press that summer, to take charge of the academic division's science and medical publishing, and there were hints that it might be possible to make my position permanent. Would I like to prepare a Publishing Plan to help the Press decide if I might be worth taking on?

On 6 November 1998 I attended a dinner party in London hosted by Penguin for Richard Dawkins to celebrate the publication of *Unweaving the Rainbow*. For me, the pre-prandial champagne was doubly appropriate: I'd been given the news earlier in the day that the Press had agreed to my being put on the payroll. From the beginning of January I would be an OUP employee.

CHAPTER TEN

SCIENTIFIC ANECDOTES
THE TEN GREAT IDEAS OF
SCIENCE

'SCIENCE WRITING AT ITS BEST'

THE FIRST SCIENTIFIC ANECDOTE I can remember hearing, recounted by my boss Bruce Wilcock during my early days at the OUP in the 1970s, describes a charming episode involving Edward Bullard. Bullard (1907–1980) was a distinguished British geophysicist and the incident took place when he was visiting an American university, on sabbatical leave. Shortly after his arrival (alas, I am not able to provide even an approximate date), Bullard was contacted by the local radio station. They liked to interview visitors to the university about their work: would he be happy to record a conversation in the studio, to be broadcast later? Afterwards, chatting to the station manager as he was being shown out of the building, Bullard asked at what time the programme would be going out on air. Astonished at the response, 2 a.m., he said he was surprised that anyone would be listening at that hour. The station manager assured him that the slot had a devoted following amongst truck drivers. By now wholly bemused, Bullard expressed amazement that truck drivers would find geophysics interesting. It was the American's turn to be taken aback. '*Geophysics*? Gee, we thought you were a palaeontologist.'

In 1975, the OUP published *The Oxford Book of Literary Anecdotes*, edited by James Sutherland, and I was in the audience when Jon Stallworthy presented it at the sales conference that summer. Jon read out selected anecdotes to whet appetites—he succeeded—and it was easy to see why the book was going to be a commercial success. The large print run and substantial book club order greatly impressed me—and made me envious—and I became convinced, there and then, that a collection of scientific anecdotes could be made just as appealing, and sell just as well.

The initial target in my sights, an editor to put together such a collection, was R. V. Jones, then Professor of Physics at the University of Aberdeen. I had listened to him at a British Association for the Advancement of Science annual meeting around that time: he seemed to carry around a vast collection of memorable stories about scientists and their milieu. I arranged to see 'Prof' in Aberdeen and visited him during subsequent editorial trips to Scotland over the next few years. He was immediately attracted to the scientific anecdotes idea but there could be no real progress until he had finished the memoir he was then writing, describing his central role in scientific military intelligence in the Second World War. This was his *Most Secret War: British Scientific Intelligence, 1939–1945*, published in 1978. I left the OUP not long after and though I stayed in touch with 'Prof', the book had been firmly fixed in my mind as the *Oxford* book of scientific anecdotes—something that seemed right only as an OUP title—so I put the project into suspended animation.

I move now to the 1980s, and my time at Longman. Tim Lincoln at *Nature* had mentioned to me that Walter Gratzer, the biophysicist and noted writer at King's College London, had been collecting material that presented diverse images of science, gathered from literature, biography, reportage, and poetry, and was looking for a publisher for this planned anthology. I lost no time in arranging to see Walter: I had met him during my brief spell at *Nature* in 1982—he came into the office once a week, as an editorial consultant—and I loved reading his book reviews, all beautifully written, all models of elegance and wit. The result was *The Longman Literary Companion to Science*, published in 1989.

My collaboration with Walter continued after my departure from Longman but the next book we worked on had not been completed until after my return to the OUP. Published in 2000, this was *The Undergrowth of Science: Delusion, Self-deception and Human Frailty*, and I felt that now was exactly the right time to revisit the original scientific anecdotes idea: *Undergrowth* and the earlier *Literary Companion to Science* seemed to be just the right starting points. As soon as *Undergrowth* had been published I asked Walter if he would take on—immediately—the scientific anecdotes. The answer was yes, and his *Eurekas and Euphorias: The Oxford Book of Scientific Anecdotes* was published in 2002. Reading batches of anecdotes as soon as they had been written was for me an intoxicating experience: I had felt confident at the outset that Walter would create something special but the creation, from its early stages, went way beyond that, and as soon as I had read the initial draft entries I became certain that the eventual book would be a winner. Even so, the sheer lavishness of the praise heaped on it by Oliver Sacks in his review in *Nature* took me by surprise. Here is Sacks getting into his stride, in the early stages of his piece: 'Gratzer has assembled a rich collection of historical incidents, conversations, fragments of autobiography and biography, and revelatory fragments, depicting moments of sudden illumination, or paralysing doubt, of intense absent-mindedness and equally intense concentration, accidents which open up wholly unexpected discoveries (there are an astonishing number of these), and incidents that cast a piercing light on the character and work of the scientist.' And finally, in full flight, his majestic climax:

> There seems to be no systematic or explicit order to the sequence of these anecdotes, but there is clearly a rightness in their ordering, and this is Gratzer the alchemist, the artist, working invisibly behind the scenes, stringing these anecdotes together like Ezra Pound's *Cantos* or the movements of an opera. Here there is tension, there relaxation; here high seriousness, there levity or farce; here the sublime, there the ridiculous. But equally one can open the book at any point and be educated, thrilled, sobered or surprised, for there is astonishment and delight on every page. . . . a banquet of epiphanies, a reference book which is also a work of art.

I had Walter Gratzer to thank for introducing me to Jim Watson, of *The Double Helix* fame. Walter and Jim had known one another since the early 1960s, when they were research scientists at Harvard. Jim was by now president of the Cold Spring Harbor Laboratory, New York (having been director from 1968 to 1993) and John Inglis, in charge of Cold Spring Harbor Laboratory Press, had proposed publishing a selection of Jim's essays. Jim liked the idea of involving a British house to co-publish the book with Cold Spring Harbor, and I certainly liked the idea of signing it up for the OUP, which I was able to arrange.

The book became *A Passion for DNA: Genes, Genomes and Society*, published in 2000. The Introduction, written by Walter Gratzer, opens thus:

> Francis Crick—the other half of the famous partnership that lit the fuse and ignited the grand new enterprise of molecular biology—recorded in his autobiography (*What Mad Pursuit*) that at the time of their great discovery his friend Jim Watson was generally regarded as too bright to be sound. There is an echo here of a sage observation, buried in Max Beerbohm's fable of Oxford life, *Zuleika Dobson*: 'the dullard's envy of brilliant men is always assuaged by the suspicion that they will come to a bad end.' But Watson and Crick have in their different ways continued to exercise a pervasive influence on the course of science, as this collection of Watson's speeches, depositions, recollections, and ruminations makes clear.

Walter's labours on the book were not yet finished. After sending his completed draft Introduction to Jim for approval, he promptly received further instructions: 'That's fine. Now please write a postscript indicating the state of molecular biology in cancer research.' The result was an Afterword, entitled 'Envoi—DNA, Peace, and Laughter'. Its opening is as delightful as that of the Introduction:

> If, dear reader, you have followed Jim Watson's discourse from his schooldays to his apotheosis as director of one of the world's great laboratories and a grandee—however puckish and irreverent—of the American scientific establishment, you will by now have a passable insight into how modern science functions and be privy also to Watson's robust view of the future.

At the end of the summer in 1999 I had arranged to visit John Inglis at Cold Spring Harbor during an editorial trip to America: some routine matters concerning *A Passion for DNA* needed to be resolved and it was convenient to do this in person whilst I was over there. John greeted me with mock formality, announcing: 'The president wishes to see you.' Jim was at that moment playing tennis, and I would be summoned as soon as he had returned to his desk. The call duly came through and John conducted me to the president's office. Jim pointed to the far end of the room. 'Michael,' he said, 'on that table is a type-script. Read as much as you can during the next hour, and then tell me over dinner what you think of it.'

The typescript was Jim's *Genes, Girls and Gamow*, and I liked the half dozen or so chapters I managed to read before it was time for dinner. I liked it even more when, the following spring, Jim sent me the complete version to read.

The autobiographical *Genes, Girls and Gamow* began where Jim's *The Double Helix* (1968), perhaps the most widely read popular science book ever published, finished: the announcement in *Nature* of Crick and Watson's proposed double-helical structure of DNA in April 1953. And the memoir ended three-and-a-half years later when Jim became a professor at Harvard. Jim's account of his life combined science, the search to understand the role of RNA in biology, with a love story: Jim's falling in love with Christa Mayr, then 17 years old, the elder daughter of the Harvard evolutionary biologist, Ernst Mayr. And woven into the story was the larger-than-life George Gamow, a Russian-born theoretical physicist, intrigued by genes, RNA, and the genetic code.

I emailed Jim as soon as I had read the typescript, saying I had found it 'charming, amusing, in places gripping, in places touching, and throughout, deeply interesting'. Here, in my view, was an important record of how things actually were, what it felt like being there, in those heroic days.

There was something else, too. In 1995, some forty years after the events described, Christa Mayr wrote to Jim. Telling him that she had kept all the letters he had sent her in the 1950s, she offered to return them. Discovering that the letters were the equivalent of entries in a meticulously kept diary, Jim used them as one of his primary sources. It gave the chronicle an immediacy, and an engaging freshness.

The book was published in 2001 and, alas, many of the reviews were unflattering. As Matt Ridley wrote (in a Foreword to *Inspiring Science, Jim Watson and the Age of DNA*, published in 2003 by Cold Spring Harbor Laboratory Press): '*Genes, Girls and Gamow* . . . broke new ground . . . abandoning smooth prose to tell a scientific saga with the clumsy style and hesitant uncertainty of a young man's mind. Had it been a novel, it would have been praised for its postmodern, innovatory syntax. But once again the world was not ready for Watson, and many literary-minded reviewers gave him no quarter.'

But one writer did like it. In a feature for a pre-Christmas issue of the *Guardian* in 2004, a range of authors selected books they had especially enjoyed that year. One of Ian McEwan's choices was *Genes, Girls and Gamow*: 'One of the strangest and most fascinating books I've read in a long while . . . ' McEwan continued:

> Once the structure of DNA was described by Crick and Watson in 1953, it took many years, and numerous scientists working in different centres, to understand the genetic code. Against this background of passionate scientific pursuit, Watson, still in his mid-20s and growing more famous by the day, fell forlornly in love with Christa, daughter of the eminent biologist, Ernst Mayr. The hopelessness of the affair is mirrored by evocations of England in the 50s—cold, inconvenient, damp. Watson was based at Caltech in California, but he toured the important labs of the world like a medieval pilgrim. Beyond love and science, there was also wilderness and hiking. The raptures, pain and restlessness of youth are finely conveyed at a breathless pace.

On 27 April 1987, BBC television transmitted *Life Story*, a dramatization of one of the great stories of science in the 20th century, the race to discover the 'secret of life': the structure of DNA, the basis of genes. It is a wonderful tale, intellectually and dramatically exciting, and the BBC rose to the occasion with a production from the top drawer. It starred Jeff Goldblum as Jim Watson, Tim Piggott-Smith as Francis Crick, Alan Howard as Maurice Wilkins, and Juliet Stevenson as Rosalind Franklin. William Nicholson wrote the screenplay, and the film was directed by Mick Jackson. I was familiar with the story— Watson and Crick working together at the Cavendish Laboratory in

Cambridge, Wilkins and Franklin, an inability to get on hampering their progress, at King's College London—having read *The Double Helix* almost twenty years earlier. Reading Jim's book had been a memorable experience; watching *Life Story* brought all of that excitement back. I remember thinking that if I had to choose one film to be shown to young people with a view to inspiring them to take up science, this would be the one.

In late 1999 I heard that Maurice Wilkins, now in his early 80s, had been working on an autobiography, and it sounded as though he had reached the stage when he would like to talk to a publisher. Maurice agreed to meet for a preliminary discussion and we had lunch at the Royal Society in London in January 2000. A synopsis and some draft chapters were subsequently prepared and sent out for review, and the proposal was formally commissioned by the OUP at the beginning of 2001. The plan was to have the book ready for publication in 2003, fifty years after the announcement in *Nature* of the structure of DNA.

The Nobel Prize awarded in 1962 for the discovery of the double helix had been shared by Watson, Crick, and Wilkins. Two of them had published their stories (Watson's *The Double Helix*, and Crick's autobiography, *What Mad Pursuit*) and the academic reviewers of the Wilkins proposal agreed on the importance of having the third laureate's version of events. There was another reason why there would be interest in the book: reading Maurice Wilkins's own account of his fraught relationship with Rosalind Franklin. Maurice explained the special importance of this in his Preface, written shortly before the book was printed. After mentioning the award of the Nobel Prize, he continued the story, in particular the consequences for him that followed the publication of *The Double Helix*:

> Some years afterwards my career took a more difficult turn. In 1968 Jim Watson wrote a book about our discovery which made some of us, Francis and me included, uncomfortable. Rosalind Franklin had died of cancer in 1958 and so she was unable to comment on Jim's book, but in 1974, the writer Anne Sayre published a book about Rosalind disputing Jim's portrayal of her. This book enabled some activists to mount a campaign in

Rosalind's name to improve the lot of women in science. This was no doubt well-intentioned and indeed useful, but one side-effect was that Rosalind's male colleagues were to some extent demonised. The most prominent demon seemed to be me. Since then, the Franklin/Wilkins story has often been told as an example of the unjustness of male scientists towards their women colleagues, and questions have been raised over whether credit was distributed fairly when the Nobel Prize was awarded. I have found this situation distressing over the years, and I expect that this book is in some way my attempt to respond to these questions, and to tell my side of that story.

Maurice didn't like the title I had chosen for the book, *The Third Man of the Double Helix*. I thought it was right (and still do) but I persuaded Maurice to accept it only at the eleventh hour. As he put it in his Preface: 'The title of this book . . . is not the one I would have chosen. I have deferred to the advice of my publishers on that issue! However, this title does resonate with some of the tensions, accusations, confusions and controversies that have attended the telling and retelling of the DNA story.'

The Third Man of the Double Helix was published in the autumn of 2003. Maurice died the following year, aged 87. Brenda Maddox, author of a fine biography of Rosalind Franklin (*Rosalind Franklin: The Dark Lady of DNA*, published in 2002), reviewed *The Third Man* in the *Times Literary Supplement*, commenting on its 'poignancy and candour'. 'And,' she continued, 'the science writing is good too, lucid to the outsider.'

I first met Charles Tanford and Jacqueline Reynolds, distinguished American scientists, retired, and living in Easingwold, North Yorkshire, towards the end of the 1980s, when I was at Longman. Enthusiastic travellers, they wanted to write a guidebook to places in Europe related to the history of scientific discovery. I listened to the case they made for publishing such a guide—tourists routinely visited art galleries and churches, but what about the scientific sites that might be just around the corner?—yet I couldn't see a market of any size for the book and so turned it down. Happily, the guide was in due course published, by Wiley. And a sequel came out a few years later. First came *The Scientific Traveler: A Guide to the People, Places and Institutions of Europe* (1992), followed by *A Travel Guide to Scientific Sites of the British Isles* (1995). I have no idea how well these books sold but they are on my shelves, and I now regret that I wasn't involved with their publication: well written, full of interest, oozing informed enthusiasm, they are splendid.

I called on Charles and Jackie again some ten years later, after my return to the OUP, and this time I didn't hesitate to express interest in their next writing project: a history of proteins. As the opening of the eventual jacket blurb proclaimed, proteins are the essence of living things. They drive our metabolism, give us immunity to disease, permit us to breathe, to move and see, and when things go wrong they can kill us. A grand subject, with a remarkable history, it embraces the origins and development of some of the key ideas of modern science, such as the possibility that giant molecules—polymers—could exist. It is a story full of extraordinary episodes and controversies, of visionary pioneers clashing with an often unreceptive and hostile establishment and—later—of the epic achievements of crystallographers, who determined the positions of all of the hundreds of thousands of atoms that make up individual protein molecules.

Charles and Jackie wanted a wide audience to enjoy their history, working hard to achieve this. Their Introduction closes with a message of encouragement: 'Our overall advice to the general reader: don't be intimidated by chemistry! Only a minimum of technical language is needed to get a feeling for the growing excitement that was generated over the years as the mysteries of protein structure and function—the core of all the mysteries of life—were revealed step by step.'

Nature's Robots: A History of Proteins had taken Charles and Jackie just over four years to write and the book was published in 2001. A review in *Nature* neatly summed up what had been achieved: 'an absorbing and often exciting story, as well as a major contribution to scholarship'.

Also published in the autumn of 2001 was Carl Djerassi's *This Man's Pill: Reflections on the 50th Birthday of the Pill*. I had met Carl the previous summer, having heard that he was writing the story of the contraceptive pill, wanting it to be published in October 2001. This would be the 50th anniversary of one of the key episodes in 20th-century social history: 15 October 1951 was the date of the first ever synthesis of an oral contraceptive. The laboratory synthesis had been carried out by Djerassi, an organic chemist, born in Vienna but educated in the United States. The event triggered the development of the Pill.

The book was short but wide ranging—from 'an account of the early history of the Pill, debunking many of the journalistic and romantic accounts of its scientific origin' to the startling revelation that the Pill was not approved in Japan until 1999 (along, in record time, with Viagra)—and all of it hugely enjoyable to read.

It was fun working with Carl, and pleasing that a high-speed production schedule resulted in the book's being ready in time for the 50th anniversary of the Pill. Shortly before publication, my colleague Kate Farquhar-Thomson, head of the academic division's publicity department at the OUP, forwarded to me an email with an accompanying comment: 'This might make you smile!' The message, forwarded to Kate by Carl, was from a journalist, and her email certainly made me smile: 'Dear Professor Djerassi, I am a feature writer for the Cyprus Mail in Cyprus. I am writing an article about the 50th anniversary of your wonderful invention. (Congratulations, and thank you!) I have comments from the Head of the Cyprus Family Planning Association ... who is a big fan of the Pill, and it would be great to have a couple of quotes from you about what you think the Pill has achieved for women. Unfortunately, the most common form of contraception here is *coitus interruptus*—need I say more. Yours sincerely . . .'

I described in Chapter 3 the experience of working on Richard Dawkins's *The Selfish Gene* and I think his book holds the record for the shortest time it took me to realize that I was on to something special: having this feeling before reaching the bottom of the first page of the typescript. The record for the *longest* time belongs to a book by Nick Lane: I had no real sense of just how exceptional it was until I read the completed typescript, three years after first hearing about the original proposal.

A letter from Nick Lane reached me at the beginning of February 1999, a few weeks after my job at the OUP had been made permanent. He wanted to write a book for a general audience and he enclosed a one-page description. The working title was *Treachery in the Air*.

My initial response, emailed to Nick, began:

Let me be open with you right at the outset. At this stage I don't know whether or not this will work as a book. To give you a feeling for why I'm saying this I need to make some general points about popular science books. In the old days people who read popular science had an earnest interest in learning more. That's no longer the case. There isn't the same amount of time available for reading and so people need to be convinced before they decide to commit to buying and reading a particular book. There is more popular science now being published: it's more competitive. What this all means is that it is no longer enough to have a book that simply explains in accessible prose an area of science: 'interesting subject', 'the author can write' are no longer sufficient. You need more!

The 'more' I was requesting included, in addition to the conventional table of contents with a paragraph summary for each chapter, a concise statement setting down what was to be unique and sellable about the book. This last had echoes for me of an OUP sales conference at the time of my early career at the beginning of the 1970s. An editor had been droning on about the forthcoming book he was presenting, was silenced in mid-sentence by the sales manager, and then given a brutally simple message: 'These reps have got ten seconds to persuade a bookseller to take this book. What should they say?' Years later, this challenging request became encapsulated in one word: hook. 'What's the hook?' became an increasingly familiar probe in publishing meetings. I explained to Nick Lane what the term meant and asked him to come up with one.

Nick in due course produced a hook—too long, but a promising start—together with a synopsis. We met for the first time in April and a couple of sample chapters were promised for the autumn: they came, they were well written, and I sent them out to academic specialists for review. The proposal was finally approved by the OUP's Delegates, and signed up, in May 2000.

There was no hint at this stage that the book would be anything more than well written but routine popular science; and this impression hadn't changed after Nick had sent me four chapters to read in February 2001. The surprise came just under a year later when I received the complete typescript. What hit me so powerfully was seeing for the first time that Nick's canvas was truly grand. Drawing on fields as diverse as geology, cosmology, chemistry, biology, and medicine, the story unravelled the unexpected ways in which oxygen spurred the evolution of life and death. Here were absorbing accounts of the origin of biological complexity, of the birth of photosynthesis, of the sudden evolution of animals, and of the need for two sexes. And the story continued right up until the present day: it offered an explanation of the accelerated ageing of cloned animals like Dolly the sheep; fresh perspectives on our own lives and deaths; insights into modern killer diseases; and an account of why we age, with accompanying advice on what we can do about it.

Nick's title when he first approached me about his proposed book was *Treachery in the Air* and this was subsequently changed, when he submitted a formal proposal, to *Life, Death and Oxygen*. It was easy to see that shortening this to *Oxygen* gave the title more punch. What was not so easy was coming up with the right subtitle and we struggled with this after the book had been completed: email exchanges over several days produced dozens of suggestions but none was quite right. Finally, Nick proposed *The Story of a Molecule that Made the World*. I loved it, and shortening it made it even better: *The Molecule that Made the World*.

Oxygen: The Molecule that Made the World was published in hardback in the autumn of 2002 and the first printing was sold out within weeks. Extracts from three reviews of the time give an impression of Nick Lane's remarkable achievement:

> Highly ambitious . . . *Oxygen* is a piece of radical scientific polemic, nothing less than a total rethink of how life evolved between about 3.5 billion and 543 million years ago, and how that relates to the diseases we suffer from today . . . This is science writing at its best.
>
> Jerome Burne, *Financial Times*

> Lane's book is an extraordinary orchestration of disparate scientific disciplines, connecting the origins of life on earth with disease, age and death in human beings.
>
> John Cornwell, *Sunday Times*

Enjoyable and informative . . . *Oxygen* presents an entertaining and cogent account of how oxidative stress fits in our rapidly expanding knowledge about ageing . . . deserves to be widely read.

Tom Kirkwood, *Nature*

I signed up Nick Lane's next book shortly after *Oxygen* was published. This was his *Power, Sex, Suicide: Mitochondria and the Meaning of Life*. Published by the OUP in 2005, a couple of years after my retirement, it was shortlisted for the Aventis (formerly Rhône-Poulenc) Science Book Prize in 2006. (One review of the book contained the following memorable comment: 'I quit smoking after reading Nick Lane's first book on oxygen, so I approached his book on sex with some trepidation.') Nick's latest book, *Life Ascending: The Ten Great Inventions of Evolution*, was published by Profile Books in 2009. It won the Science Book Prize (now sponsored by the Royal Society) in 2010.

Nick Lane's original book proposal, *Treachery in the Air*, eventually becoming *Oxygen*, came to me in a way that I particularly liked: it was Nick's first book and he had contacted me on the recommendation of one of my authors (John Emsley). I was almost invariably out of the running when an author had become an established success, with a proposal arriving via an agent, because I wasn't working for the sort of publisher accustomed to offering substantial royalty advances. Another way of finding new authors wanting to write their first book was coming across them unexpectedly. Here is one such story, with its pleasing conclusion.

At the beginning of 2002 I was asked to represent the OUP on the judging panel for a competition jointly sponsored by what was then the *Times Higher Education Supplement* and the OUP, the THES/OUP Science Essay competition. It was open to anyone, and the first prize was £2,000, plus the *Oxford English Dictionary* on CD-ROM. The shortlisting, in February, was the responsibility of Martin Ince, deputy editor of THES, and me: we read the sixty or so entries, came up with our own lists, and after an exchange of emails eventually agreed on a shortlist of 12. In March, Martin and I joined three professors at the offices of THES and spent the morning choosing the best three essays, ranking them for first, second, and third prize.

A favourite entry had emerged for me during the earlier sifting process. Its subject was diet and dyslexia: an element of surprise at the outset, with a proposed link between two topics I hadn't previously thought had any connection. The piece was readable and clear, the basic minimum to be expected in an essay competition, but what singled it out was that it told an elaborate story with consummate skill. The areas of science involved, from neuroscience to the properties of cell membranes, are not especially easy to explain to the outsider, but the author had managed it beautifully.

The diet and dyslexia essay won first prize. Only then, after we had reached our decision, did we discover the identity of the author (during judging, each essay had been anonymized and given a code number): her name was Kathleen Taylor, a postdoctoral scientist in the physiology department at Oxford University. With a subeditor's enticing headline, 'A recipe for healthy brain growth: start with fish oil' and snappy introduction, 'Dyslexia can ruin lives. Now evidence suggests that the right diet may diminish or even prevent this and other neuronal disorders . . . ', the prize-winning essay was published in the THES at the end of March. For good measure, six months later, in October 2002, Kathleen won the THES Humanities and Social Sciences Writing Prize.

I met Kathleen for the first time in the middle of April, when she came into the OUP offices to collect her prize. A small group of us had gathered to welcome her, and we drank some champagne. Kathleen was initially noticeably quiet but she became more animated when I asked her if she had given any thought to writing a book. 'Yes,' was the immediate reply. 'Do you have a particular subject in mind?' There was another instant response: 'Brainwashing.'

One of the high spots in my life as a commissioning editor was hearing for the first time about a juicy idea for a book. This was one of those moments: coming out of the blue, here was an intriguing subject, and more, there was an instinctive feeling that a book likely to sell was in prospect.

Kathleen set down the background in a formal proposal, which reached me three weeks later. The term 'brainwashing', although first recorded in 1950, is an expression of a much older concept: the forcible, full-scale alteration of a target person's beliefs. It gradually became a familiar part of popular culture; and a subject too of learned discussion from a variety of angles, from history and sociology to psychology and psychotherapy, and, not least, marketing. But one aspect had been noticeably missing from all of these discussions: any serious reference to real brains. Descriptions of how opinions can be changed—by persuasion, deceit, or force—had been almost entirely psychological. Neuroscientists, meanwhile, had been working to unravel the mind's workings, drawing on information provided by social psychology. Currently, Kathleen pointed out in her proposal, these two disciplines —neuroscience and social psychology—rarely communicated. In her book, she proposed to bring them together.

Brainwashing: The Science of Thought Control was published in 2004. Here is one reaction, a short review in *Focus*, the BBC's science and technology magazine, published at the end of 2004:

> A magisterially detailed survey . . . Taylor is never less than direct and engaging. The subject may be difficult but the writing never is. With no hint at all of academic pretension, this is a model of how to make hard science accessible without rendering it impossibly watered down or patronising. This is an outstanding book. Academic researchers and human rights professionals will find it a goldmine of relevant research and information. And anyone else interested in psychology will find it a thrill.

It is time to tell the last of my tales, concluding with the book I link with *The Selfish Gene* in my own book's subtitle, Peter Atkins's *Galileo's Finger*.

For the first part of my spell at the OUP that followed my change in status from freelance to employee, I worked for two separate departments, one concerned with the college market (university-level textbooks), the other with trade (general reader) books. On the college side, I became responsible for the famously successful textbooks of Peter Atkins. I had commissioned his *Physical Chemistry* during the 1970s (described in Chapter 4) and now found myself overseeing its seventh edition, then being written, together with a clutch of other texts carrying the Atkins name. Once a year I would attend a meeting of the American Chemical Society, during which Peter and I had meetings with Jessica Fiorillo, W. H. Freeman's chemistry editor in New York, to settle numerous issues related to the various new editions on which Peter was engaged, published jointly by the OUP and W. H. Freeman. On one of these occasions, a working session with Jessica, Peter accused me of not being sufficiently interested in his new editions, because 'your heart is in trade'. I protested my innocence, though I secretly knew he was right. Developing and publishing new editions of successful textbooks is a hugely important business, yet it wasn't something that I found wholly fulfilling. Textbooks were fine, provided they were new, rather than new editions. But my heart was indeed by then in trade.

Back in Oxford, wearing my college hat, I had monthly working lunches with Peter to review progress with his textbooks. The meeting on 3 May 2001 proceeded normally until the end, when Peter announced that he had suddenly decided he wanted to write a trade book. It was to be about the ten seminal ideas of modern science. Spellbound, I listened as he reeled off examples of the ideas he had in mind. 'I love it,' I said, after he'd finished. 'I knew you'd say that,' came the reply. Then I spelled out the problem. If he took the proposal to an agent, rival publishers would doubtless offer royalty advances way beyond what the OUP was used to paying. I said that I really wanted the book. What should we do? Peter replied that he didn't know, though said he would very much like to work with me on the book. Then came the central message: if the Press wanted to win first-division books, it would need to review its thinking on advances. A formal publishing proposal—a synopsis—would be ready in a couple of weeks' time and I undertook to see how far the OUP could be persuaded to go in order to secure this special book.

The advance we eventually offered at the end of 2001 was, for the OUP, handsome. Peter acknowledged this, and accepted it. By this time he had come up with a title, *Galileo's Finger*, with *The Ten Great Ideas of Science* as the subtitle.

Peter delivered each draft chapter on its completion, for comments from me on behalf of the general reader, and for sending out to academic specialists to vet the science. A short, handwritten note accompanied the last chapter. Dated 6 May 2002, it read: 'M, Well—here we are. Another . . .' Below, displayed as an illustration, was a full stop. It was a lovely echo of a note Peter had sent me in April 1976 (see Chapter 4, page 67) with its 'almost exact facsimile of the full stop at the end of the final chapter of the little thing I am doing for you. The original was created on Tuesday evening . . .'

In September I prepared for sales and marketing colleagues a note to give a feeling for why *Galileo's Finger* was special. The note comprised ten points, matching the book's subtitle, though that was a coincidence. Here is one of the points:

It might be thought that the book—the elegant writing and the author's expository gifts accepted—is actually really no more than an overview of major ideas from the past and from contemporary science: things already known, with nothing new. I'd like to respond to this with an anecdote. Peter and I met regularly during the writing. On one of these occasions he told me about the chapter on quantum theory that he was then finishing. He approached each chapter in the same way: first reading (widely and deeply) until his understanding was such that he was confident he could write about the subject for his intended general audience. He and I had both earlier predicted that the quantum theory chapter would be reasonably straightforward: the subject was for him very much home territory. But this didn't turn out to be the case. He discovered that not a single book explained the central mystery of quantum theory. Eventually, after a mighty intellectual struggle, it—the explanation of the mystery—came to him. This is real creativity. There have been during the writing a few other epiphanies: not a huge number, not on every page, not in every chapter, but there are things in this book that are genuine breakthroughs in explaining the profound. It takes me back 25 years, when the great W. D. Hamilton told me that critics of Dawkins's *The Selfish Gene*, who said that it merely explained what was already known, had failed to recognize that the book contained a number of genuinely new ideas.

Galileo's Finger was published in March 2003. Richard Dawkins had agreed to read it in proof and we printed his endorsement on the jacket: 'The Nobel Prize for Literature has never been won by a scientist. It is high time it happened, and Peter Atkins would be my candidate. . . . he uses his powerful mastery of the English language to open our eyes to the poetry of deep science. . . . [his] literate prose leaves us inspired, fulfilled, enriched, and properly alive.'

On 31 March 2003, Kate Farquhar-Thomson, head of the publicity department of the OUP's academic division, circulated an email to everyone in the division. With the subject heading TOP TEN BEST-SELLER, it read: 'As far as I am aware, the OUP Trade & Reference Department have their first top ten bestseller since 1998 [the *New Oxford Dictionary of English*]: *Galileo's Finger* by Peter Atkins. Number 10, *Sunday Times*, Hardbacks (General), 30 March 2003. I am thrilled!'

If, dear reader—to recycle Walter Gratzer's words at the beginning of his Afterword for Jim Watson's *A Passion for DNA*, quoted earlier in this chapter—you have reached this far, I hope you will have gained some insight into publishing and the life it brings. But much more important, my hope is that you will have had your appetite whetted. That is, that you will have felt motivated to want to read at least one of the books described. Might they really be as appealing as my descriptions made out? I think they might!

EPILOGUE

I RETIRED from the OUP towards the end of 2003, but it was several years before the idea of attempting to write about publishing gradually took hold of me. There were some interesting tales to tell, or at least I convinced myself that there were, but this was not the principal reason for wanting to try my hand as an author. The real motivation came from wanting to communicate to others something of the intellectual excitement I'd experienced 'meeting and working with extraordinary people, and discovering at first hand ideas that, when hearing about them for the first time, were simply breathtaking', as I put it towards the end of my Prologue. I wanted to open windows and let the ideas themselves seduce the general reader, especially one with no background in science. For me, it would be wonderful if reading just one of my stories ignited enough curiosity to motivate someone to go out and buy and read the book in question.

The point has occasionally been made, to illustrate a fundamental difference between the arts and science, that if Shakespeare hadn't lived, we wouldn't have Shakespeare's plays; but we *would* have the theory of relativity had Einstein not lived: someone else would have formulated it, eventually. Richard Dawkins's *The Selfish Gene* and Peter Atkins's *Galileo's Finger*, to take the two books that inspired this volume's subtitle, would have been published had I not been around as a commissioning editor. But I *was* around, and being involved with them was hugely enjoyable; and a privilege.

An important point to be made about working on those two books, and many others, is that I was given the editorial freedom to pursue them and then work with their authors. (I commented in Chapter 6 on the reasons behind this editorial freedom in the context of my working as a commissioning editor at W. H. Freeman, and in Chapter 2 in connection with my job as a field editor at the OUP.) Times gradually changed during the course of my career, with the introduction in particular of appraisals and targets. Longman introduced a bonus system when I worked there in the 1980s. It combined the company's and the division's performance (actual results compared with the original budget) and personal performance (in my own case, how many books I had signed up in the year compared with a target, set in an appraisal meeting a year earlier). This system could, and did, result in the payment of a substantial bonus. W. H. Freeman, on the other hand, both in America and in the UK, did not have a bonus system, certainly not for commissioning editors, because it was recognized that bonuses tend to lead to indifferent commissioning: mediocre books are easier to sign up, and meet targets, than first-rate ones. W. H. Freeman's policy was the right one. So my view on managing commissioning editors, once they have demonstrated that they can do the job, is to give them plenty of editorial freedom. But I acknowledge that this is easier said than done!

'Teach yourself publishing' is a fair description of how things were for me in the early stages of my career. Those days are of course long gone, and there is no better illustration of this than that provided by the growth of university publishing courses. A three-year Diploma in Publishing at the Oxford College of Technology in 1961 was the first full-time publishing qualification to be offered in British higher education. The first undergraduate degree course in publishing began in 1982 at Oxford Polytechnic (which became Oxford Brookes University in 1992), a pioneering development reported by the *Observer* newspaper. Publishing, the *Observer* piece commented, was a very tough trade. 'Too tough', it went on, 'for the gentlemanly amateur with nothing but flair.' The new degree course at Oxford Polytechnic offered to equip its graduates with 'specialized skills as well as general publishing knowhow'. The writer in the *Observer* ventured: 'Dare one say that by 1985, when the first degrees are handed out, England will be producing her very first fully qualified publishers?'

Today, about a dozen universities in the UK offer undergraduate or postgraduate courses on publishing (Oxford Brookes offers both). Is it now possible to get a job in publishing without such a qualification? I am told that it is, but that it is a great deal easier with one!

In 2010 I gave a talk to the publishing postgraduate class at Oxford Brookes, repeating it in 2011 and 2012. Throughout my career as a commissioning editor in science, I had always associated postgraduate classes with small numbers, this explaining my surprise when I gave my talks at Brookes. The class each time contained around ninety students, for me a huge number. Something else struck me too: the proportion of women. Of the ninety in each class, only half a dozen or so were men. This ratio, I learned, closely matched the one in the book publishing industry as a whole.

There is another asymmetry that intrigues me. During my time at Longman in the 1980s the science group occasionally needed to recruit a desk editor. This was a recognized entry level into publishing for graduates. Working with a commissioning editor, at Longman called a publisher, the editor was responsible for overseeing a book's progress during the editing and production stages, from delivered typescript to printed books. Advertisements were placed for a desk editor, the basic qualification being a science degree, and we typically received fewer than ten applications. When my colleagues in the arts group advertised for a desk editor with an arts degree they were invariably rewarded with close to a hundred applications.

Let me now return to my talks to the postgraduate classes at Brookes. On one occasion I asked for a show of hands: how many in the class of ninety had a science degree? Three or four hands went up.

It would please me greatly if reading this book persuaded just one science graduate to go into publishing . . .

Have I missed publishing and the life it brings? I thought of this a few years after my retirement when I was being entertained to lunch by one of my former authors. If only he had the time, he said, he would love to write a book he had in mind. Then he went on to describe it. Suddenly, I recognized the symptoms: the mounting excitement on listening to an author describing a book they'd like to write, feeling certain that it would work, and sell. That is one thing I miss.

NOTES AND REFERENCES

FOREWORD

The quoted extract is taken from Richard Dawkins, 'Growing up in ethology', in *Leaders in Animal Behavior: The Second Generation*, edited by Lee Drickamer and Donald Dewsbury, Cambridge University Press, 2010, Chapter 8, p. 207.

CHAPTER ONE

Jeremy Bernstein's review of *A Brief History of Time* by Stephen Hawking: 'A Brief History of Time', *New Yorker*, 6 June 1988, pp. 117–122.

Michael Rodgers, 'The Hawking phenomenon', *Public Understanding of Science*, Volume 1, Issue 2, April 1992, pp. 231–234.

John Manger's piece on the review coverage of Pais's biography of Einstein: 'Study in indifference', *Nature*, Volume 324, 13 November 1986, p. 169.

J. G. Crowther, *Fifty Years with Science*, Barrie & Jenkins, London, 1970, p. 67.

Albert Einstein, 'Einstein on Newton', *Manchester Guardian*, 19 March 1927, pp. 13–14.

CHAPTER TWO

J. G. Crowther, *Fifty Years with Science*, Barrie & Jenkins, London, 1970. For Crowther's recollections on C. P. Scott, see p. 49; for Eddington and Rutherford, see pp. 37–38; and for Dirac, see p. 39.

Letters to *The Times*, May 1972, from S. Jacobson ('Soaring price of books', 15 May, p. 15), L. D. Barron ('The price of books', 18 May, p. 19), C. H. Roberts ('The price of books today', 22 May, p. 15).

CHAPTER THREE

R. C. Lewontin's review of *The Selfish Gene* by Richard Dawkins: 'Caricature of Darwinism', *Nature*, Volume 266, 17 March 1977, pp. 283–284.

Letter from W. D. Hamilton: 'The Selfish Gene', *Nature*, Volume 267, 12 May 1977, p. 102.

(This elicited a response from Lewontin: 'The Selfish Gene', *Nature*, Volume 267, 19 May 1977, p. 202.)

CHAPTER FOUR

Michael Rodgers's review of *Wiley: One Hundred and Seventy Five Years of Publishing* by John Hammond Moore: 'Ruskin then, science now', *Nature*, Volume 300, 4 November 1982, p. 87.

Michael Rodgers's post-obituary letter about Lord Dainton: *Independent*, 22 December 1997, p. 16.

CHAPTER SIX

The Nobel laureates listed on the W. H. Freeman tee shirt were Harold Varmus (Physiology or Medicine, 1989), Leon Lederman (Physics, 1988), David Hubel (Physiology or Medicine, 1981), Steven Weinberg (Physics, 1979), Ilya Prigogine (Chemistry, 1977), Julian Schwinger (Physics, 1965), Max Perutz (Chemistry, 1962), James Watson (Physiology or Medicine, 1962), Arthur Kornberg (Physiology or Medicine, 1959), and Linus Pauling (Chemistry, 1954 and Peace, 1962).

Eleanor Lawrence's *A Guide to Modern Biology: Genetics, Cells and Systems*, Longman Scientific & Technical, Harlow, 1989.

CHAPTER EIGHT

There are telling but affectionate descriptions of Hamilton's shortcomings as a lecturer in Ullica Segerstråle's biography, *Nature's Oracle: the life and work of W. D. Hamilton* (Oxford University Press, 2013). They include an account from R. Kitching, an undergraduate at the time of Hamilton's first lecture at Imperial College (p. 113), and one from R. L. Trivers describing a lecture by Hamilton to faculty and students at Harvard (p. 137).

CHAPTER NINE

Andrew Boa's review of *Organic Chemistry* by Jonathan Clayden, Nick Greeves, Stuart Warren, and Peter Wothers: 'Names throw their weight around in O-chem', *Times Higher Education Supplement* (now the *Times Higher Education*), 2 March 2001, reviews section, pp. XVIII– XIX.

CHAPTER TEN

Oliver Sacks's review of *Eurekas and Euphorias: The Oxford Book of Scientific Anecdotes* by Walter Gratzer: 'Bringing scientists to life', *Nature*, Volume 419, 24 October 2002, p. 786.

Kathleen Taylor's prize-winning essay on diet and dyslexia, 'A recipe for healthy brain growth: start with fish oil', was published in the *Times Higher Education Supplement*, 29 March 2002, pp. 16–17. Her other essay published later that year, winner of the THES/Palgrave Humanities and Social Sciences Writing Prize, entitled 'Is imagination more important than knowledge', was published under the headline 'When fact and fantasy collide' in the *Times Higher Education Supplement*, 20 December 2002, p. VIII (internal section between pp. 18 and 19).

EPILOGUE

Piece on the first undergraduate degree course in publishing, 'A degree of practicality' by Michael Dineen, *Observer*, 24 May 1981, p. 21.

INDEX